SUBMARINE EXPOSURE GUIDANCE LEVELS FOR SELECTED HYDROFLUOROCARBONS

HFC-236fa, HFC-23, and HFC-404a

Subcommittee on Exposure Guidance Levels
for Selected Hydrofluorocarbons

Committee on Toxicology

Board on Environmental Studies and Toxicology

Commission on Life Sciences

National Research Council

NATIONAL ACADEMY PRESS
Washington, D.C.

NATIONAL ACADEMY PRESS   2101 Constitution Ave., N.W.   Washington, D.C. 20418

NOTICE: The project that is the subject of this report was approved by the Governing Board of the National Research Council, whose members are drawn from the councils of the National Academy of Sciences, the National Academy of Engineering, and the Institute of Medicine. The members of the committee responsible for the report were chosen for their special competences and with regard for appropriate balance.

This project was supported by Contract Nos. DAMD 17-89-C-9086 and DAMD 17-99-C-9049 between the National Academy of Sciences and the U.S. Department of Defense. Any opinions, findings, conclusions, or recommendations expressed in this publication are those of the author(s) and do not necessarily reflect the view of the organizations or agencies that provided support for this project.

International Standard Book Number 0-309-07084-8

Additional copies of this report are available from:

National Academy Press
2101 Constitution Ave., NW
Box 285
Washington, DC 20055

800-624-6242
202-334-3313 (in the Washington metropolitan area)
http://www.nap.edu

Copyright 2000 by the National Academy of Sciences. All rights reserved.

Printed in the United States of America

THE NATIONAL ACADEMIES

National Academy of Sciences
National Academy of Engineering
Institute of Medicine
National Research Council

The **National Academy of Sciences** is a private, nonprofit, self-perpetuating society of distinguished scholars engaged in scientific and engineering research, dedicated to the furtherance of science and technology and to their use for the general welfare. Upon the authority of the charter granted to it by the Congress in 1863, the Academy has a mandate that requires it to advise the federal government on scientific and technical matters. Dr. Bruce M. Alberts is president of the National Academy of Sciences.

The **National Academy of Engineering** was established in 1964, under the charter of the National Academy of Sciences, as a parallel organization of outstanding engineers. It is autonomous in its administration and in the selection of its members, sharing with the National Academy of Sciences the responsibility for advising the federal government. The National Academy of Engineering also sponsors engineering programs aimed at meeting national needs, encourages education and research, and recognizes the superior achievements of engineers. Dr. William A. Wulf is president of the National Academy of Engineering.

The **Institute of Medicine** was established in 1970 by the National Academy of Sciences to secure the services of eminent members of appropriate professions in the examination of policy matters pertaining to the health of the public. The Institute acts under the responsibility given to the National Academy of Sciences by its congressional charter to be an adviser to the federal government and, upon its own initiative, to identify issues of medical care, research, and education. Dr. Kenneth I. Shine is president of the Institute of Medicine.

The **National Research Council** was organized by the National Academy of Sciences in 1916 to associate the broad community of science and technology with the Academy's purposes of furthering knowledge and advising the federal government. Functioning in accordance with general policies determined by the Academy, the Council has become the principal operating agency of both the National Academy of Sciences and the National Academy of Engineering in providing services to the government, the public, and the scientific and engineering communities. The Council is administered jointly by both Academies and the Institute of Medicine. Dr. Bruce M. Alberts and Dr. William A. Wulf are chairman and vice chairman, respectively, of the National Research Council.

SUBCOMMITTEE ON EXPOSURE GUIDANCE LEVELS FOR SELECTED HYDROFLUOROCARBONS

GARY CARLSON *(Chair)*, Purdue University, West Lafayette, Indiana
MARION W. ANDERS, University of Rochester, Rochester, New York
DAROL E. DODD, ManTech Environmental Technology, Inc., Dayton, Ohio
HARIHARA M. MEHENDALE, Northeast Louisiana University, Monroe, Louisiana
CHARLES REINHARDT, Chadds Ford, Pennsylvania
ANNETTE SHIPP, The K.S. Crump Group, Inc., Ruston, Louisiana
MARY VORE, University of Kentucky, Lexington, Kentucky
ROBERT YOUNG, Oak Ridge National Laboratory, Oak Ridge, Tennessee

Staff

SUSAN N.J. PANG, Project Director
RUTH E. CROSSGROVE, Editor
MIRSADA KARALIC-LONCAREVIC, Information Specialist
LEAH L. PROBST, Project Assistant
EMILY L. SMAIL, Project Assistant

Sponsor
U.S. NAVY

COMMITTEE ON TOXICOLOGY

BAILUS WALKER, JR. *(Chair)*, Howard University Medical Center, Washington, D.C.
MELVIN E. ANDERSEN, Colorado State University, Denver, Colorado
GERMAINE M. BUCK, State University of New York at Buffalo
GARY P. CARLSON, Purdue University, West Lafayette, Indiana
JACK H. DEAN, Sanofi Pharmaceuticals, Inc., Malverne, Pennsylvania
ROBERT E. FORSTER II, University of Pennsylvania, Philadelphia, Pennsylvania
PAUL M.D. FOSTER, Chemical Industry Institute of Toxicology, Research Triangle Park, North Carolina
DAVID W. GAYLOR, U.S. Food and Drug Administration, Jefferson, Arkansas
JUDITH A. GRAHAM, U.S. Environmental Protection Agency, Research Triangle Park, North Carolina
SIDNEY GREEN, Howard University, Washington, D.C.
WILLIAM E. HALPERIN, National Institute for Occupational Safety and Health, Cincinnati, Ohio
CHARLES H. HOBBS, Lovelace Respiratory Research Institute and Lovelace Biomedical and Environmental Research Institute, Albuquerque, New Mexico
FLORENCE K. KINOSHITA, Hercules Incorporated, Wilmington, Delaware
MICHAEL J. KOSNETT, University of Colorado Health Sciences Center, Denver, Colorado
MORTON LIPPMANN, New York University School of Medicine, Tuxedo, New York
THOMAS E. MCKONE, University of California, Berkeley, California
ERNEST E. MCCONNELL, ToxPath, Inc., Raleigh, North Carolina
DAVID H. MOORE, Battelle Memorial Institute, Bel Air, Maryland
GÜNTER OBERDÖRSTER, University of Rochester, Rochester, New York
JOHN L. O'DONOGHUE, Eastman Kodak Company, Rochester, New York
GEORGE M. RUSCH, AlliedSignal, Inc., Morristown, New Jersey
MARY E. VORE, University of Kentucky, Lexington, Kentucky
ANNETTA P. WATSON, Oak Ridge National Laboratory, Oak Ridge, Tennessee

Staff

KULBIR S. BAKSHI, Program Director
SUSAN N.J. PANG, Program Officer
ABIGAIL STACK, Program Officer
RUTH E. CROSSGROVE, Publications Manager
KATHRINE J. IVERSON, Manager, Toxicology Information Center
EMILY L. SMAIL, Project Assistant

BOARD ON ENVIRONMENTAL STUDIES AND TOXICOLOGY

GORDON ORIANS (*Chair*), University of Washington, Seattle, Washington
DONALD MATTISON (*Vice Chair*), March of Dimes, White Plains, New York
DAVID ALLEN, University of Texas, Austin, Texas
INGRID C. BURKE, Colorado State University, Fort Collins, Colorado
WILLIAM L. CHAMEIDES, Georgia Institute of Technology, Atlanta, Georgia
JOHN DOULL, University of Kansas Medical Center, Kansas City, Kansas
CHRISTOPHER B. FIELD, Carnegie Institute of Washington, Stanford, California
JOHN GERHART, University of California, Berkeley, California
J. PAUL GILMAN, Celera Genomics, Rockville, Maryland
BRUCE D. HAMMOCK, University of California, Davis, California
MARK HARWELL, University of Miami, Miami, Florida
ROGENE HENDERSON, Lovelace Respiratory Research Institute, Albuquerque, New Mexico
CAROL HENRY, Chemical Manufacturers Association, Arlington, Virginia
BARBARA HULKA, University of North Carolina, Chapel Hill, North Carolina
JAMES F. KITCHELL, University of Wisconsin, Madison, Wisconsin
DANIEL KREWSKI, University of Ottawa, Ottawa, Ontario
JAMES A. MACMAHON, Utah State University, Logan, Utah
MARIO J. MOLINA, Massachusetts Institute of Technology, Cambridge, Massachusetts
CHARLES O'MELIA, Johns Hopkins University, Baltimore, Maryland
WILLEM F. PASSCHIER, Health Council of the Netherlands
KIRK SMITH, University of California, Berkeley, California
MARGARET STRAND, Oppenheimer Wolff Donnelly & Bayh, LLP, Washington, D.C.
TERRY F. YOSIE, Chemical Manufacturers Association, Arlington, Virginia

Senior Staff

JAMES J. REISA, Director
DAVID J. POLICANSKY, Associate Director and Senior Program Director for Applied Ecology
CAROL A. MACZKA, Senior Program Director for Toxicology and Risk Assessment
RAYMOND A. WASSEL, Senior Program Director for Environmental Sciences and Engineering
KULBIR BAKSHI, Program Director for the Committee on Toxicology
LEE R. PAULSON, Program Director for Resource Management
ROBERTA M. WEDGE, Program Director for Risk Analysis

COMMISSION ON LIFE SCIENCES

MICHAEL T. CLEGG *(Chair)*, University of California, Riverside, California
PAUL BERG *(Vice Chair)*, Stanford University, Stanford, California
FREDERICK R. ANDERSON, Cadwalader, Wickersham & Taft, Washington, D.C.
JOANNA BURGER, Rutgers University, Piscataway, New Jersey
JAMES E. CLEAVER, University of California, San Francisco, California
DAVID EISENBERG, University of California, Los Angeles, California
JOHN EMMERSON, Fishers, Indiana
NEAL FIRST, University of Wisconsin, Madison, Wisconsin
DAVID J. GALAS, Keck Graduate Institute of Applied Life Science, Claremont, California
DAVID V. GOEDDEL, Tularik, Inc., South San Francisco, California
ARTURO GOMEZ-POMPA, University of California, Riverside, California
COREY S. GOODMAN, University of California, Berkeley, California
JON W. GORDON, Mount Sinai School of Medicine, New York, New York
DAVID G. HOEL, Medical University of South Carolina, Charleston, South Carolina
BARBARA S. HULKA, University of North Carolina, Chapel Hill, North Carolina
CYNTHIA KENYON, University of California, San Francisco, California
BRUCE R. LEVIN, Emory University, Atlanta, Georgia
DAVID LIVINGSTON, Dana-Farber Cancer Institute, Boston, Massachusetts
DONALD R. MATTISON, March of Dimes, White Plains, New York
ELLIOT M. MEYEROWITZ, California Institute of Technology, Pasadena, California
ROBERT T. PAINE, University of Washington, Seattle, Washington
RONALD R. SEDEROFF, North Carolina State University, Raleigh, North Carolina
ROBERT R. SOKAL, State University of New York, Stony Brook, New York
CHARLES F. STEVENS, The Salk Institute, La Jolla, California
SHIRLEY M. TILGHMAN, Princeton University, Princeton, New Jersey
RAYMOND L. WHITE, University of Utah, Salt Lake City, Utah

Staff

WARREN R. MUIR, Executive Director
JACQUELINE K. PRINCE, Financial Officer
BARBARA B. SMITH, Administrative Associate
LAURA T. HOLLIDAY, Senior Program Assistant

OTHER REPORTS OF THE
BOARD ON ENVIRONMENTAL STUDIES AND TOXICOLOGY

Copper in Drinking Water (2000)
Ecological Indicators for the Nation (2000)
Waste Incineration and Public Health (1999)
Hormonally Active Agents in the Environment (1999)
Research Priorities for Airborne Particulate Matter: I. Immediate Priorities and a Long-Range Research Portfolio (1998); II. Evaluating Research Progress and Updating the Portfolio (1999)
Ozone-Forming Potential of Reformulated Gasoline (1999)
Risk-Based Waste Classification in California (1999)
Arsenic in Drinking Water (1999)
Brucellosis in the Greater Yellowstone Area (1998)
The National Research Council's Committee on Toxicology: The First 50 Years (1997)
Toxicologic Assessment of the Army's Zinc Cadmium Sulfide Dispersion Tests (1997)
Carcinogens and Anticarcinogens in the Human Diet (1996)
Upstream: Salmon and Society in the Pacific Northwest (1996)
Science and the Endangered Species Act (1995)
Wetlands: Characteristics and Boundaries (1995)
Biologic Markers (5 reports, 1989-1995)
Review of EPA's Environmental Monitoring and Assessment Program (3 reports, 1994-1995)
Science and Judgment in Risk Assessment (1994)
Ranking Hazardous Waste Sites for Remedial Action (1994)
Pesticides in the Diets of Infants and Children (1993)
Issues in Risk Assessment (1993)
Setting Priorities for Land Conservation (1993)
Protecting Visibility in National Parks and Wilderness Areas (1993)
Dolphins and the Tuna Industry (1992)
Hazardous Materials on the Public Lands (1992)
Science and the National Parks (1992)
Animals as Sentinels of Environmental Health Hazards (1991)
Assessment of the U.S. Outer Continental Shelf Environmental Studies Program, Volumes I-IV (1991-1993)
Human Exposure Assessment for Airborne Pollutants (1991)
Monitoring Human Tissues for Toxic Substances (1991)
Rethinking the Ozone Problem in Urban and Regional Air Pollution (1991)
Decline of the Sea Turtles (1990)

Copies of these reports may be ordered from
the National Academy Press
(800) 624-6242 or (202) 334-3313
www.nap.edu

OTHER REPORTS OF THE COMMITTEE ON TOXICOLOGY

Toxicological Risks of Selected Flame-Retardant Chemicals (2000)
Arsenic in Drinking Water (1999)
Review of the U.S. Army's Health Risk Assessments for Oral Exposure to Six Chemical-Warfare Agents (1999)
Toxicity of Military Smokes and Obscurants, Volume 1(1997), Volume 2 (1999), Volume 3 (1999)
Assessment of Exposure-Response Functions for Rocket-Emission Toxicants (1998)
Review of a Screening Level Risk Assessment for the Naval Air Facility at Atsugi, Japan (Letter Report) (1998)
Toxicologic Assessment of the Army's Zinc Cadmium Sulfide Dispersion Tests (1997)
Toxicologic Assessment of the Army's Zinc Cadmium Sulfide Dispersion Tests: Answers to Commonly Asked Questions (1997)
Toxicity of Alternatives to Chlorofluorocarbons: HFC-134a and HCFC-123 (1996)
Permissible Exposure Levels for Selected Military Fuel Vapors (1996)
Spacecraft Maximum Allowable Concentrations for Selected Airborne Contaminants, Volume 1 (1994), Volume 2 (1996), Volume 3 (1996), Volume 4 (2000)

Preface

AS PART of the effort to phase out the use of stratospheric ozone-depleting substances, such as chlorofluorocarbons (CFCs), the U.S. Navy is considering hydrofluorocarbons (HFCs) as replacements for the CFC refrigerants used aboard its submarines. Before using the HFCs, the Navy plans to set emergency exposure guidance levels (EEGLs) and continuous exposure guidance levels (CEGLs) to protect submariners from health effects that could occur as a result of accidental releases or slow leaks.

In this report, the Subcommittee on Exposure Guidance Levels for Selected Hydrofluorocarbons of the National Research Council's (NRC's) Committee on Toxicology independently reviews the scientific validity of the Navy's proposed 1-hr and 24-hr EEGLs and 90-day CEGLs for two of the candidate refrigerants—HFC-236fa and HFC-404a. In addition, the subcommittee reviews the the EEGLs and CEGL for HFC-23, one of the combustion products of HFC-236fa. This NRC report is intended to aid the Navy in using HFCs safely.

This report has been reviewed in draft form by individuals chosen for their technical expertise and diverse perspectives in accordance with procedures approved by the NRC's Report Review Committee for reviewing NRC and Institute of Medicine reports. The purpose of that independent review was to provide candid and critical comments to assist the NRC in making the published report as sound as possible and to ensure that the report meets institutional standards for objectivity, evidence, and responsiveness to the study charge. The review comments and draft manuscript remain confidential to protect the integrity of the deliberative process. We wish to thank the following individuals, who are neither officials nor employees of the NRC, for their participation in the review of this report: Melvin Andersen, Colorado State University; John Doull, The University of Kansas Medical Center; Ian Greaves, University of Minnesota; Robert Hamlin, Ohio State University; Joseph Rodricks, The Life Sciences Consultancy; and Richard Schlesinger, New York University School of Medicine.

The individuals listed above have provided many constructive comments and suggestions. It must be emphasized, however, that responsibility for the final content of this report rests entirely with the authoring committee and the NRC.

We gratefully acknowledge Dr. William Brock of DuPont's Haskell Laboratory and Dr. George Rusch of AlliedSignal, Inc., for providing background information on the HFCs and for making presentations to the subcommittee.

We are grateful for the assistance of the NRC staff in preparing the report. Staff members who contributed to this effort are Carol A. Maczka, senior program director for the Toxicology and Risk Assessment Program; Kulbir S. Bakshi, senior program director of the Committee on Toxicology; Ruth E. Crossgrove, editor; Leah Probst and Emily Smail, project assistants; and Mirsada Karalic-Longcarevic, information specialist. We especially wish to recognize the contributions of project director, Susan N.J. Pang, who coordinated the project and contributed to the preparation of the subcommittee's report.

Finally, we would like to thank all the members of the subcommittee for their dedicated efforts throughout the development of this report.

Gary P. Carlson, Ph.D.
Chair, Subcommittee on Exposure Guidance Levels for Selected Hydrofluorocarbons

Bailus Walker, Jr., Ph.D., M.P.H.
Chair, Committee on Toxicology

Contents

SUMMARY 1

1 INTRODUCTION 9
   Statement of Task, 10
   Approach to the Study, 10
   Exposure Guidance Levels, 10
   Structure of the Report, 13
   References, 13

2 HYDROFLUOROCARBON-236FA 14
   Chemical and Physical Properties, 14
   Toxicokinetics, 15
   Toxicity Information, 16
   Summary, 21
   Exposure Guidance Levels, 22
   Recommendations, 25
   References, 26

3 HYDROFLUOROCARBON-23 27
   Chemical and Physical Properties, 27
   Toxicokinetics, 28
   Toxicity Information, 29
   Summary, 34

Exposure Guidance Levels, **36**
References, **37**

4 HYDROFLUOROCARBON-404A **40**
HFC-143a, **41**
HFC-125, **47**
HFC-134a, **54**
Summary, **71**
Exposure Guidance Levels for HFC-404a, **73**
Calculations, **74**
References, **76**

Submarine Exposure Guidance Levels For Selected Hydrofluorocarbons: HFC-236fa, HFC-23, and HFC-404a

Summary

IN ACCORDANCE WITH the Montreal Protocol on Substances That Deplete the Ozone Layer, which calls for phasing out the use of chlorofluorocarbons (CFCs), the U.S. Navy proposes to replace CFC refrigerants aboard its submarines with hydrofluorocarbons HFC-236fa and HFC-404a. To protect submariners from adverse health effects resulting from exposure to accidental releases of those compounds, the Navy plans to set emergency exposure guidance levels (EEGLs) and continuous exposure guidance levels (CEGLs) for them.

An EEGL is defined as a concentration of a substance in air that is judged to be acceptable for the performance of specific tasks during rare emergency conditions lasting for periods of 1-24 hr. EEGLs are intended to prevent irreversible harm and degradation in crew performance. Temporary discomfort, such as eye or upper-respiratory-tract irritation, is permissible as long as there is no effect on judgment, performance, or ability to respond to an emergency.

To protect submariners from exposures that are longer than 24 hr, CEGLs are set to provide a ceiling guidance level for up to 90 days of continuous exposure to a substance. The intent of a CEGL is to avoid any adverse health effects, either immediate or delayed, associated with prolonged exposures and to avoid any degradation in performance. Some conditions, such as a slight headache, which might be acceptable for short periods in emergency situations, are not permissible for longer-term exposures.

The Navy proposes to use the same exposure guidance levels for HFC-

236fa and HFC-404a that it established for two chlorofluorocarbons (CFC-12 and CFC-114): a 1-hr EEGL of 2,000 parts per million (ppm), a 24-hr EEGL of 1,000 ppm, and a 90-day CEGL of 100 ppm. The Navy also proposes to apply those guidance levels to HFC-23, a combustion product of HFC-236fa. HFC-23 is formed when escaped HFC-236fa is passed through a submarine's carbon monoxide and hydrogen burners.

STATEMENT OF TASK

The National Research Council (NRC) was asked to conduct an independent evaluation of the Navy's proposed exposure guidance levels for HFC-236fa, HFC-23, and HFC-404a. The NRC assigned this task to the Committee on Toxicology (COT), which convened the Subcommittee on Exposure Guidance Levels for Selected Hydrofluorocarbons, a multidisciplinary group of experts. The subcommittee was asked to review the available toxicity data on the three HFCs and to determine the scientific validity of the Navy's proposed EEGLs and CEGLs. The evaluation was to include an assessment of the relevance of animals studies for evaluating risks to humans, the completeness of the data base, the target organs of toxicity, and the appropriateness of the methods used to derive the guidance levels (e.g., correctly adjusting for exposure durations and the use of uncertainty factors). The subcommittee was asked to review the three HFCs only in the context of use aboard submarines—vessels with male personnel only. The subcommittee was also asked to identify any deficiencies in the data base on each HFC and to make recommendations for future research. Although products other than HFC-23 are formed during combustion of HFC-236fa or HFC-404a, HFC-23 was the only combustion product the subcommittee was asked to consider.

APPROACH TO THE STUDY

The subcommittee conducted a critical analysis of the available toxicity data on each of the HFCs and used the data to calculate possible EEGLs and CEGLs according to the guidelines outlined in the NRC's 1986 report *Criteria and Methods of Preparing Emergency Exposure Guidance Level (EEGL), Short-Term Public Emergency Guidance Level (SPEGL), and Continuous Exposure Guidance Level (CEGL) Documents*. The subcommittee also reviewed the Navy's toxicity assessments of the HFCs and used information provided in a 1996 report of the NRC titled *Toxicity of Alternatives to Chlorofluorocarbons: HFC-134a and HCFC-123*, which provides a toxicity assessment and establishes EEGLs and CEGLs for HFC-134a, a component of HFC-404a

CONCLUSIONS AND RECOMMENDATIONS

Table S-1 presents a comparison of the EEGLs and CEGLs recommended by the subcommittee with those proposed by the Navy. The subcommittee concludes that the guidance levels proposed by the Navy are unnecessarily conservative. In all cases, the subcommittee's recommended levels are greater than those proposed by the Navy. The reason for the difference is that the Navy did not use data on the HFCs to calculate the proposed exposure guidance levels, but rather proposed to use the same guidance levels established for chlorofluorocarbons CFC-12 and CFC-114. The subcommittee believes there are adequate data on the individual HFCs to calculate scientifically valid exposure guidance levels.

TABLE S-1 Submarine Exposure Guidance Levels

Exposure	NRC Levels,[a] ppm	Navy Levels,[b] ppm
HFC-236fa		
1-hr EEGL	10,000	2,000
24-hr EEGL	2,000	1,000
90-day CEGL	350	100
HFC-23		
1-hr EEGL	20,000	2,000
24-hr EEGL	5,000	1,000
90-day CEGL	500	100
HFC-404a		
1-hr EEGL	12,900	2,000
24-hr EEGL	4,300	1,000
90-day CEGL	800	100

[a]Calculated on the basis of the available data.
[b]The proposed exposure guidance levels are the same as those previously established for chlorofluorocarbons CFC-12 and CFC-114.

HFC-236fa

On the basis of the available data on HFC-236fa, the subcommittee recommends a 1-hr EEGL of 10,000 ppm, a 24-hr EEGL of 2,000 ppm, and a 90-day CEGL of 350 ppm. The 1-hr EEGL is based on a cardiac sensitization study in which the no-observed-adverse-effect level (NOAEL) for HFC-236fa in dogs was 100,000 ppm for a 5-min exposure. The NOAEL was divided by an uncertainty factor of 10 to account for interspecies variability, yielding a value of 10,000 ppm for exposures up to 1 hr.

A 14-week toxicity study in rats was used to determine the 24-hr EEGL for HFC-236fa. In the study, a NOAEL of 20,000 ppm was identified on the basis of decrements in "alerting response" (the response to a sudden auditory stimulus). The subcommittee had reservations about using alerting response as a toxicity end point, because effects were transient, responses were subjectively evaluated (and are known to vary among strains and individual animals), and it is unclear whether such effects are applicable to humans. However, in the absence of other data regarding effects caused by HFC-236fa, alerting response was considered to be the most appropriate available end point. A 24-hr EEGL of 2,000 ppm was determined by dividing the NOAEL of 20,000 ppm by an uncertainty factor of 10 to account for interspecies variability.

The 14-week toxicity study in rats was also considered to be the most relevant study for calculating the 90-day CEGL. The NOAEL of 20,000 ppm was divided by an uncertainty factor of 10 to extrapolate from animals to humans to yield a value of 2,000 ppm. That value was adjusted to account for the discontinuous exposure regimen used in the study by multiplying 2,000 ppm by 1/4 (to account for exposure for 6 hr per day) and by 5/7 (to account for exposure five times per week), which yielded a 90-day CEGL of 350 ppm.

If HFC-236fa is considered for use on vessels with female crew members, the 24-hr EEGL and 90-day CEGL might have to be reconsidered on the basis of maternal toxicity. A developmental toxicity study in rats reported reduced weight gain and decrements in alerting response in pregnant animals exposed at 20,000 ppm. That is the same concentration as the NOAEL in the 14-week toxicity study used to calculate the 24-hr EEGL and the 90-day CEGL.

Uncertainties exist with regard to the effects that HFC-236fa might have on human performance. End points of narcosis and decrements in alerting response have been observed in laboratory animals, but it is unclear whether human performance would be similarly affected. Because HFC-236fa was relatively nontoxic in laboratory studies and because similar HFCs, such as HFC-23 and HFC-134a, have been shown to have low toxicity in humans, the subcommittee recommends that tests be conducted with humans to determine whether HFC-236fa affects performance skills, such as motor coordination and alertness.

HFC-23

The subcommittee believes that data on HFC-23 support a 1-hr EEGL of 20,000 ppm, a 24-hr EEGL of 5,000 ppm, and a 90-day CEGL of 500 ppm. The basis for the 1-hr EEGL was a human exposure study in which subjects were intermittently exposed to HFC-23 (eight exposures of 3 min each with 2-min intervals of exposure to air only). The NOAEL for the study was 200,000 ppm. To account for the discontinuous exposure, the NOAEL was divided by a factor of 10, resulting in a 1-hr EEGL of 20,000 ppm. For the 24-hr EEGL, a developmental toxicity study was used. Although such a developmental study is not necessarily the most appropriate study for deriving a 24-hr exposure guidance level, when considering the all-male population aboard submarines, that study had the most relevant exposure duration (a total of 90 hr), and no maternal or fetal effects were observed at the highest dose tested of 50,000 ppm. The NOAEL was divided by a factor of 10 to account for interspecies differences, resulting in an exposure guidance level of 5,000 ppm. The basis for the 90-day CEGL of 500 ppm was a 90-day continuous exposure study in dogs, in which the NOAEL was 5,000 ppm. That value was divided by an uncertainty factor of 10 to account for interspecies variability.

HFC-404a

HFC-404a is a gaseous mixture of three halocarbons—52% HFC-143a, 44% HFC-125, and 4% HFC-134a. The subcommittee believes that the most appropriate way to calculate exposure guidance levels for HFC-404a is the method used by the American Conference of Governmental Industrial Hygienists to calculate threshold limit values for special cases when the exposure of concern is a liquid mixture and the atmospheric composition is assumed to be similar to that of the original material (i.e., on a time-weighted-average exposure basis, all the liquid mixture eventually evaporates). In this case, when the percent composition by weight of the liquid mixture is known, the exposure guidance levels (EGLs) can be determined using the following equation:

$$\text{EGL of mixture} = \frac{1}{\dfrac{f_a}{EGL_a} + \dfrac{f_b}{EGL_b} + \dfrac{f_c}{EGL_c} + \ldots \dfrac{f_n}{EGL_n}}$$

where f is the fraction of each particular component. The component's corresponding EGL is expressed in units of milligrams per cubic meter (mg/m^3). To use this equation, it was necessary to calculate the exposure guidance levels for the individual components of HFC-404a.

Exposure Guidance Levels for Components of HFC-404a

The three halocarbons that comprise HFC-404a are HFC-143a, HFC-125, and HFC-134a. For component HFC-143a, the subcommittee estimated a 1-hr EEGL of 25,000 ppm on the basis of a cardiac sensitization study in dogs. The NOAEL for the study was 250,000 ppm, and an uncertainty factor of 10 was applied to account for interspecies variability. For the 24-hr EEGL, a 4-week toxicity study in rats was used; the highest tested concentration of 40,000 ppm was the NOAEL. To extrapolate from animals to humans, the NOAEL was divided by an uncertainty factor of 10, yielding a 24-hr EEGL of 4,000 ppm. A 90-day toxicity study in rats was used to calculate the 90-day CEGL for HFC-143a. The NOAEL for the study was 40,000 ppm, which was divided by 10 to account for interspecies variability; the resulting value of 4,000 ppm was adjusted to account for the study's discontinuous exposure regimen by multiplying it by 1/4 (to account for exposure for 6 hr per day) and by 5/7 (to account for exposure five times per week), which yielded a 90-day CEGL of 700 ppm.

For component HFC-125, a cardiac sensitization study in dogs was used to derive a 1-hr EEGL. The NOAEL for the study was 75,000 ppm, and an uncertainty factor of 10 was applied to account for interspecies variability, yielding a 1-hr EEGL of 7,500 ppm. The subcommittee calculated a 24-hr EEGL of 5,000 ppm on the basis of a 4-week toxicity study in rats. The highest tested concentration of 50,000 ppm was the NOAEL, and that value was divided by an uncertainty factor of 10 to account for interspecies differences. For the 90-day CEGL, a 90-day toxicity study in rats was used. The NOAEL for that study was 50,000 ppm, and an uncertainty factor of 10 was applied for to account for interspecies differences. To account for the discontinuous exposure regimen used in the study, 5,000 ppm was multiplied by 1/4 (to account for exposure for 6 hr per day) and by 5/7 (to account for exposure five times per week), resulting in a 90-day CEGL of 900 ppm.

In 1996, the COT reviewed the available toxicity data on HFC-134a and proposed a 1-hr EEGL of 4,000 ppm, a 24-hr EEGL of 1,000 ppm, and a 90-day CEGL of 900 ppm. Since that review, additional data on HFC-134 have become available. One of the new studies was an ascending-concentration safety study in humans, in which subjects were exposed to HFC-134a at concentrations up to 8,000 ppm for 1 hr with no adverse effects. The subcommittee believes that, on the basis of that study, a 1-hr EEGL of 8,000 ppm for HFC-134a is justified. For the 24-hr EEGL, the subcommittee used a 13-week toxicity study in rats, in which the highest concentration tested of 50,000 ppm was the NOAEL. Dividing the NOAEL by an uncertainty factor of 10 to account for interspecies variability yielded a 24-hr EEGL of 5,000 ppm. That exposure level is higher than the EEGL of 1,000 ppm recommended by the NRC in 1996. The reason for the difference is that in 1996 the NRC was determining exposure levels for use aboard Navy ships with female crew members and, therefore, based the NOAEL of 10,000 ppm on a developmental study in which fetal toxicity was observed. However, fetal toxicity is not as a relevant an end point for setting an exposure level for use on submarines, which have no female crew members. For the 90-day toxicity study, the subcommittee agreed with the NRC's earlier proposal of 900 ppm. That exposure level was based on a 2-year toxicity study, in which the NOAEL was 50,000 ppm. That value was divided by an uncertainty factor of 10 and then adjusted for the discontinuous exposure regimen used in the study by multiplying it by 1/4 (to account for exposure 6 hr per day) and by 5/7 (to account for exposure five times per week).

Exposure Guidance Levels for HFC-404a

Using the equation presented earlier and the exposure levels calculated above for the individual components of HFC-404a, the 1-hr EEGL, 24-hr EEGL, and 90-day CEGL for HFC-404a were calculated to be 12,900 ppm, 4,300 ppm, and 800 ppm, respectively.

1

Introduction

IN ACCORDANCE WITH the Montreal Protocol on Substances That Deplete the Ozone Layer, efforts are under way to replace the chlorofluorocarbons (CFCs) used in refrigeration units. One class of chemicals replacing the CFCs is the hydrofluorocarbons (HFCs). These chemicals do not contain chlorine and do not contribute to the destruction of the stratospheric ozone layer. Their use as replacements for the CFCs in refrigeration units is expected to expand.

The U.S. Navy proposes to use HFC-236fa and HFC-404a as refrigerants aboard its submarines; specifically, HFC-236fa is being considered for use in centrifugal chillers, and HFC-404a is being considered for use in ice-cream machines. Because of the closed environment of submarines, the Navy plans to set emergency exposure guidance levels (EEGLs) and continuous exposure guidance levels (CEGLs) to protect its personnel from potential adverse health effects, both short- and long-term, caused by inhalation of those chemicals as the result of accidental releases.

The Navy proposes to set a 1-hr EEGL of 2,000 parts per million (ppm), a 24-hr EEGL of 1,000 ppm, and a 90-day CEGL of 100 ppm for HFC-236fa and HFC-404a. The Navy also proposes to apply those guidance levels to HFC-23, a combustion product of HFC-236fa. These levels are the same as those established for chlorofluorocarbons CFC-12 and CFC-114.

STATEMENT OF TASK

The National Research Council (NRC) was asked to conduct an independent evaluation of the Navy's proposed exposure guidance levels for HFC-236fa, HFC-23, and HFC-404a. The NRC assigned this task to the Committee on Toxicology (COT), which convened the Subcommittee on Exposure Guidance Levels for Selected Hydrofluorocarbons, a multidisciplinary group of experts, to review (1) the Navy's toxicity assessments of the three HFCs, (2) other data relevant to establishing EEGLs and CEGLs for these substances, and (3) the scientific validity of the Navy's proposed EEGLs and CEGLs, based on the relevance of existing animal studies for evaluating the health risks to humans, the completeness of the data base, the target organs of toxicity, and the appropriateness of the methods used to derive the exposure guidance levels (e.g., correctly adjusting for exposure durations and the use of uncertainty factors). The subcommittee was asked to review the three HFCs only in the context of use aboard submarines—vessels with male personnel only. The subcommittee was also asked to identify any deficiencies in the data bases on each HFC and to make recommendations for future research. Although products other than HFC-23 are formed during combustion of HFC-236fa or HFC-404a, HFC-23 was the only combustion product the subcommittee was asked to consider.

APPROACH TO THE STUDY

The subcommittee conducted a critical analysis of the available toxicity data on each of the HFCs and used the data to calculate possible EEGLs and CEGLs according to the guidelines outlined in the NRC's report *Criteria and Methods of Preparing Emergency Exposure Guidance Level (EEGL), Short-Term Public Emergency Guidance Level (SPEGL), and Continuous Exposure Guidance Level (CEGL) Documents* (NRC 1986). The subcommittee also reviewed the Navy's toxicity assessments of the HFCs and used information provided in a 1996 report of the NRC titled *Toxicity of Alternatives to Chlorofluorocarbons: HFC-134a and HCFC-123* (NRC 1996), which provides a toxicity assessment and establishes EEGLs and CEGLs for HFC-134a, a component of HFC-404a.

EXPOSURE GUIDANCE LEVELS

Criteria and methods used to determine EEGLs and CEGLs are detailed in an NRC (1986) report, and a brief description is provided below. Additional guidance is provided in two other NRC reports (NRC 1992, 1993).

Emergency Exposure Guidance Levels

An EEGL is defined as a ceiling guidance level for single emergency exposures usually lasting from 1 to 24 hr—an occurrence expected to be infrequent in the lifetime of a person. "Emergency" connotes a rare and unexpected situation with potential for significant loss of life, property, or mission accomplishment if not controlled. An EEGL can also be set for much shorter periods, such as 1-min or 5-min exposures. An EEGL specifies and reflects the subcommittee's interpretation of available information in the context of an emergency.

An EEGL is acceptable only in an emergency, when some risks or some discomfort must be endured to prevent greater risks (such as fire, explosion, or massive release). Even in an emergency, exposure should be limited to a defined short period. Exposure at the EEGL might produce temporary discomfort, such as eye or upper-respiratory-tract irritation, headache, or increased respiratory rate. The EEGL is intended to prevent irreversible harm. Even though some reduction in performance is permissible, it should not prevent proper responses to the emergency (such as shutting off a valve, closing a hatch, removing a source of heat or ignition, or using a fire extinguisher). For example, in normal work situations, a degree of upper-respiratory-tract irritation or eye irritation causing discomfort would not be considered acceptable; during an emergency, however, such irritation would be acceptable if it did not cause irreversible harm or seriously affect judgment or performance. The EEGL for a substance represents the subcommittee's judgment based on evaluation of experimental and epidemiological data, mechanisms of injury, and, when possible, operation conditions in which emergency exposure might occur, as well as consideration of DOD goals and objectives.

The EEGL is generally calculated on the basis of acute toxicity end points. However, even brief exposure to some substances might have the potential to increase the risk of cancer or other delayed effects. If the substance under consideration is carcinogenic, a cancer risk assessment is performed with the aim of providing an estimate of the exposure that would not lead to an excess risk of cancer greater than 1 in 10,000 exposed persons. The acceptable risk selected for military exposures is based on considerations of policy and objectives of DOD.

In estimating the EEGL for a substance that has multiple biological effects, all end points—including respiratory, neurological, reproductive (in both sexes), developmental, carcinogenic, and other organ-related effects—are evaluated, and the most important is selected. If confidence in the available data is low or if important data are missing, appropriate uncertainty factors are used and the rationale for their selection is stated. Generally, EEGLs have been developed for exposure to single substances, although emergency exposures often involve complex mixtures of substances and thus have a potential for toxic synergism. In the absence of other information, guidance levels for complex mixtures can be developed from EEGLs by assuming as a first approximation that the toxic effects are simply additive—thus implying a proportional reduction in EEGLs for each of the constituents of a mixture.

Continuous Exposure Guidance Levels

The CEGL is a ceiling guidance level set to prevent adverse health effects, either immediate or delayed, of prolonged exposures and to prevent degradation in crew performance that might endanger the objectives of a particular mission as a consequence of continuous exposure for up to 90 days. In contrast with EEGLs, CEGLs are intended to provide guidance for exposures under what is expected to be normal operating conditions in a submarine for periods of up to 90 days. Some conditions, such as slight headache, which might be acceptable for short periods under emergency conditions would not be permissible for long-term exposures. Because long-term exposures are repeated or continuous, detoxification and excretion are of special importance as they relate to the potential of the chemical to accumulate in the body.

Special Considerations

One important consideration that is commonly used in establishing exposure guidance levels for the general public is variability among humans in sensitivity to the effects of chemicals. A default uncertainty factor of 10 is typically used to protect susceptible individuals. However, the subcommittee believes that using an uncertainty factor to account for intraspecies variability is not necessary because the submariner population is all male, young (18-30 years of age), and healthier than the general population, having passed rigorous physical and psychological examinations.

Another consideration in the derivation of the EEGLs and CEGLs is the inhalation toxicokinetics of the HFCs. In general, the uptake of an inhaled HFC is a function of the rate of respiration (pulmonary ventilation), solubility of the HFC in the blood (blood:gas partition coefficient), pulmonary blood flow (cardiac output), and partial pressure of HFC in the blood. Because of the marked difference in pulmonary ventilation between rodents and humans, rodents will reach a constant HFC arterial pressure much more rapidly than will humans. These differences could be important for short-term exposures, but would likely be of less importance for long-term exposures where both humans and rodents would ultimately reach steady-state arterial concentrations of the gases.

STRUCTURE OF THE REPORT

The results of the subcommittee's evaluation of HFC-236fa, HFC-23, and HFC-404a are presented in Chapters 2, 3, and 4, respectively. For each agent, the subcommittee evaluates inhalation data on the toxicokinetics; acute, subchronic, and chronic toxicity; reproductive effects; developmental effects; genotoxicity; and carcinogenic effects. In addition, special consideration is given to data on cardiac sensitization because inhalation of HFCs and similar compounds are known to make the mammalian heart abnormally sensitive to epinephrine, resulting in cardiac arrhythmia and possibly death.

REFERENCES

NRC (National Research Council). 1986. Criteria and Methods of Preparing Emergency Exposure Guidance Level (EEGL), Short-Term Public Emergency Guidance Level (SPEGL), and Continuous Exposure Guidance Level (CEGL) Documents. Washington D.C: National Academy Press.

NRC (National Research Council). 1992. Guidelines for Developing Spacecraft Maximum Allowable Concentrations for Space Station Contaminants. Washington D.C: National Academy Press.

NRC (National Research Council). 1993. Guidelines for Developing Community Emergency Exposure Levels for Hazardous Substances. Washington D.C: National Academy Press.

NRC (National Research Council). 1996. Toxicity of Alternatives to Chlorofluorocarbons: HFC-134a and HCFC-123. Washington D.C: National Academy Press.

2

Hydrofluorocarbon-236fa

H<small>YDROFLUOROCARBON</small> (HFC)-236<small>FA</small>, or 1,1,1,3,3,3-hexafluoropropane, is a gaseous halocarbon that is being considered as a replacement for the refrigerant chlorofluorocarbon (CFC)-114, which is used in centrifugal chillers aboard U.S. Navy submarines. To protect submariners from large accidental releases or low-level slow leaks of HFC-236fa, emergency exposure guidance levels (EEGLs) and continuous exposure guidance levels (CEGLs) are needed to avoid adverse health effects from short-term or prolonged exposures to HFC-236fa and to avoid degradation in crew performance. This chapter presents the available toxicity information on HFC-236fa and the subcommittee's evaluation of the Navy's proposed 1-hr and 24-hr EEGLs and 90-day CEGL.

CHEMICAL AND PHYSICAL PROPERTIES

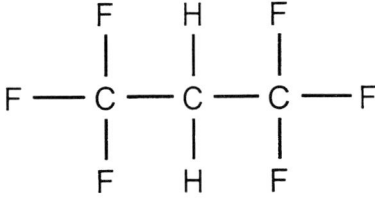

Common name: HFC-236fa
Chemical name: 1,1,1,3,3,3-hexafluoropropane
Synonyms: Hexafluoropropane; 2,2-Dihydrohexafluoropropane
CAS number: 690-39-1
Structural formula: $CF_3CH_2CF_3$
Description: Colorless gas
Molecular weight: 152.01
Boiling point: $-0.7°C$
Melting point: $-93.6°C$
Density and specific gravity: 1.370 g/cc
Vapor pressure: 36 psia at 25°C (calculated)
Conversion factors: 1 mg/m^3 = 0.16 ppm; 1 ppm = 6.22 mg/m^3

TOXICOKINETICS

Using tissues obtained from male rats, Vinegar et al. (1995) determined tissue and air partition coefficients for HCF-236fa by vial equilibration. Coefficients were determined for blood/air, liver/air, fat/air, gut/air, rapidly perfused tissue/air, and slowly perfused tissue/air in incubations of 3 hr at 37°C with 800 parts per million (ppm) of HFC-236fa. The partition coefficients (mean ± standard deviation of 12 determinations) were found to be 0.49 ± 0.04 (blood/air), 0.56 ± 0.06 (liver/air), 3.69 ± 0.56 (fat/air), 0.56 ± 0.06 (gut/air), 0.56 ± 0.06 (rapidly perfused tissues/air), and 0.87 ± 0.08 (slowly perfused tissues/air).

Gas-uptake experiments were performed by Vinegar et al. (1995) by exposing three male rats for 6 hr to HFC-236fa via inhalation at concentrations of 100, 530, 2350, 7300, and 18,000 ppm. Loss runs, tests performed without rats to determine the loss rate of HFC-236fa from the chamber, showed percent losses of 0.38% ± 0.07% to 2.85% ± 1.40% per hr, and loss runs with animals in the chamber ranged from 1.57% to 11.77%. For both situations, loss of HFC-236fa from exposure chambers was greatest at lower concentrations. Humidity levels, initially thought to be affecting loss rates, were found to have no appreciable effect on the loss of the test material. The role of carbon dioxide in the inexplicable loss of test material was considered but not investigated. Inhalation uptake of HFC-236fa by rats was biphasic with a rapid equilibration phase of up to 30 min followed by a slow linear uptake phase. The partition-coefficient data and the data obtained from the gas-uptake experiments were used in a physiologically based pharmacokinetic (PB-PK) model in an attempt to describe mathematically the disposition and metabolism of HFC-236fa. The PB-PK model was unable to describe adequately the loss of the test material from the animal chamber. Data from the gas-uptake experiments were inconsistent with metabolism-mediated disappearance of HFC-236fa.

Samples of blood, urine, or feces were collected from rats exposed to HFC-236fa after 6 and 24 hr of exposure (Vinegar et al. 1995). The samples were extracted with hexane or cyclohexane and analyzed by gas chromatography and mass spectrometry (GC-MS); some samples were methylated before GC-MS analysis to detect organic acids. Although HFC-236fa was detected in samples of blood, urine, or feces, no fluorocarbon metabolites of HFC-236fa were detected by GC-MS, either in the total-ion-current mode or in the single-ion-monitoring mode. Moreover, GC-MS analysis revealed no compounds with retention times consistent with seven fluorocarbons proposed as possible metabolites of HFC-236fa.

Valentine (1995) found no fluoride ions in the urine of rats exposed to HFC-236fa concentrations as high as 50,000 ppm for 6 hr per day, 5 days per week for 2 weeks, thereby indicating no significant metabolism of the compound.

TOXICITY INFORMATION

Acute Toxicity

Currently available data indicate that HFC-236fa has low acute toxicity by the inhalation route. Keller (1994) exposed young male rats to concentrations of HFC-236fa at 150,000 or 200,000 ppm (purity 99.06%) for 4 hr. Actual mean concentrations were approximately 134,000 and 189,000 ppm, respectively. During the whole-body exposure, the oxygen concentration was maintained at 21% ± 3% and air flow was approximately 2 liters (L) per min with a total chamber volume of 13 L. No rats died during exposure or during an additional day of observation following the exposure. Exposure to HFC-236fa at 134,000 ppm produced no observable effects, but rats exposed at 189,000 ppm exhibited narcosis (nonresponsive to sound) that persisted for approximately 30 min after cessation of exposure. No additional effects were observed during the 1-day post-exposure period, and no pathological evaluations were conducted. Under the conditions of this study, the 4-hr no-observed-adverse-effect level (NOAEL) for narcosis in rats is at or above 134,000 ppm, and the 4-hr lowest-observed-adverse-effect level (LOAEL) for narcosis is at or below 189,000 ppm.

In an acute inhalation study by Ulrich (1996), a group of five male and five female young adult rats was exposed (whole body) to HFC-236fa (purity >99.5%) at a nominal concentration of 471,000 ppm (actual mean concentration was 457,000 ppm) for 4 hr. Chamber oxygen was maintained at 21.5% ± 5.33% with an air flow of 18.7 L/min. Although chamber temperature ranged from 19.7 to 30.8°C, there was no apparent effect on the test animals. Shortly after initial exposure, the rats exhibited a brief period of hyperactivity before becoming prostrate. No toxicologically significant effects were observed 1 hr after cessation of exposure or during the 14-day post-exposure observation period. A transient loss of body weight was observed in two female rats (3 and 9 g, respectively, during the first 3 days after exposure), but terminal weights exceeded those of day 0. Necropsy revealed dark red lungs in one male and three females, renal cysts in two females, and an enlarged pituitary in one female. The pathological findings were not considered to be exposure related. On the basis of the results of this study, 457,000 ppm is considered a 4-hr NOAEL for HFC-236fa inhalation exposure in rats.

Cardiac Sensitization

Huntington Research Centre (HRC 1994) examined the cardiac sensitization potential of HFC-236fa in six young adult male beagle dogs. The specific epinephrine dose to elicit a minimal response, as determined by altered electrocardiogram (ECG) with a few ectopic beats, was determined for each dog and found to range from 2 to 12 μg/kg. Dogs were exposed by face mask to HFC-236fa at concentrations of 50,000, 100,000, 150,000, 200,000, 250,000, or 300,000 ppm. An epinephrine challenge injection was administered after 5 min of exposure, and ECG monitoring was continued for another 5 min after the challenge. Evidence for cardiac sensitization to epinephrine challenge was observed in two of six dogs following exposure to HFC-236fa at 150,000 ppm as shown by multifocal ventricular ectopic activity in one dog and a fatal ventricular fibrillation in another. Following a 5-min exposure of six dogs to 200,000 ppm, two of the dogs (one of which also exhibited a positive response at 150,000 ppm) exhibited multifocal ventricular ectopic activity. At higher concentrations, excitation or narcosis in the dogs prevented full evaluation of the cardiac effects of HFC-236fa. The results of this study provided evidence that HFC-236fa might cause cardiac sensitization (including a fatality) in dogs following 5-min exposures to concentrations at or above 150,000 ppm. Under the conditions of this study, 100,000 ppm may be considered a NOAEL.

Subchronic Toxicity

Available data indicate that HFC-236fa is also of low toxicity following intermittent, subchronic exposure.

Smith et al. (1997) reported on a subchronic toxicity study in which groups of five male and five female rats were exposed (nose only) to HFC-236fa at 50,000 ppm for 6 hr per day for 5 consecutive days. The rats were observed for an additional 14 days. Although some rats reportedly exhibited signs of respiratory-tract irritation, no gross clinical abnormalities or gross pathological correlates were observed. Under the conditions of this study, the 50,000-ppm exposure may be considered a NOAEL.

In a 2-week whole-body inhalation exposure study, groups of five male and five female young adult rats were exposed to HFC-236fa (purity >99.9%) at target concentrations of 0, 5,000, 20,000, or 50,000 ppm for 6 hr per day, 5 days per week for 2 weeks (Valentine 1995). The control group was exposed to clean air only. Mean airflow in the 350 L chamber was approximately 64 L/min over the duration of the experiment, and oxygen was maintained at approximately 19%. Temperature and relative humidity were maintained at acceptable levels. Analytical concentrations varied only slightly (±10%) from the target values. No rats died during the course of the exposure, and there were no significant changes in body weight that could be attributed to the HFC exposure. Alerting response (response to a sudden auditory stimulus) was abolished or diminished in all the rats in the 50,000-ppm group on test days 1 and 2. By test days 3 and 4, alerting responses had returned to normal or near-normal in these rats. Throughout the remainder of the test period, rats in the 50,000-ppm group exhibited only transient decrements in alerting responses (observed in only one of three observation intervals for each exposure).

With the exception of transient decrements in the 20,000-ppm group on test days 1 and 2, normal alerting responses were observed in all the rats in all the groups during the actual exposure period. With the exception of one rat in the highest-exposure group, normal alerting response was observed at 30-50 min after the daily exposure, indicating that any decrements observed were quickly reversible, although considered exposure related. On the basis of these findings, the NOAEL for decrement in alerting response was 5,000 ppm. Clinical pathological evaluations revealed no exposure-related, toxicologically relevant findings immediately after the last exposure. A 60% decrease in hepatic β-oxidation was detected in male rats in the 50,000-ppm group, but there were no gross or histopathological correlates suggesting peroxisome proliferation. Under the conditions of this study, 5,000 ppm may be considered a NOAEL and 20,000 ppm a LOAEL; the latter is based on diminution of alerting response.

Valentine (1996) also reported a 90-day whole-body inhalation study of HFC-236fa. In this study, groups of 10 male and 10 female rats were exposed to HFC-236fa at nominal concentrations of 0, 5,000, 20,000, or 50,000 ppm (analytical concentrations were 0, 4,980, 20,000, and 50,300 ppm, respectively) for 6 hr per day, 5 days per week for 14 weeks. Two additional groups of 10 male rats were added on day 50 for assessment of peroxisomal and mitochondrial β-oxidation activity. One group served as a control group, and the other was exposed as previously described to HFC-236fa at 50,000 ppm for 2 weeks. Five rats from each of these two groups were sacrificed immediately after exposure, and all remaining rats were sacrificed after a 2-week post-exposure period. Environmental and air-flow conditions in the 350-L chamber were as described for the 2-week study.

Although minor excursions of temperature and humidity beyond the desired ranges occurred, they were not great as to compromise the validity of the study results. No rats died as a result of exposure to the test article, and there were no significant and sustained changes in body weights of exposed rats. Although varied, nonspecific clinical signs (alopecia, ocular and nasal discharges, and staining) were observed in some rats in most groups immediately before or after the exposure sessions, these observations were of inconsistent occurrence and considered to be spurious. Rats exposed at 5,000 or 20,000 ppm exhibited normal alerting responses. On study day 1, most rats in the 50,000-ppm group exhibited a diminished response.

During the first 2 weeks of the study, one to three rats in the 50,000-ppm group exhibited a diminished alerting response that was generally limited to the last 2 hr of the daily exposure period, and that was reversible within 30 to 50 min following cessation of exposure. By study day 18, rats in the 50,000-ppm group were no longer exhibiting any alterations in alerting response.

Clinical chemistry evaluations revealed significant decreases in serum cholesterol of males in the 50,000-ppm group and significant decreases in serum triglycerides in males in all the exposure groups and in females in the

20,000 and 50,000-ppm groups. Although considered exposure related, these changes are not considered to be of toxicological significance. Decreases in total protein and albumin and alterations in blood urea nitrogen (BUN) concentration were also detected but were not dose related. Changes in some serum electrolyte concentrations were not considered to be biologically relevant. Urinalysis revealed no biologically relevant findings. There was no definitive evidence for induction of hepatic peroxisomes in rats following discontinuous subchronic exposure to HFC-236fa at concentrations as high as 50,000 ppm.

Reproductive Toxicity

There are no reports of reproductive toxicity studies in animals for HFC-236fa.

Developmental Toxicity

Two developmental toxicity studies (one in rats and one in rabbits) were conducted in animals to examine the developmental effects of HFC-236fa. The highest concentration tested in these studies—50,000 ppm—was the highest that could be attained without supplementing chamber oxygen.

Munley (1995) administered HFC-236fa by inhalation to pregnant rats at concentrations of 0, 5,000, 20,000 and 50,000 ppm for 6 hr per day from days 7 to 16 of gestation. The study was terminated on day 22 of gestation. At 20,000 and 50,000 ppm, there were significant dose-related decreases in maternal body-weight gain over the first 2 days of inhalation exposures. At 50,000 ppm, that was accompanied by significant reduction in food consumption and diminished alerting responses during the inhalation exposures. No evidence of maternal toxicity was detected at 5,000 ppm. There was no evidence of developmental toxicity in the fetuses at any exposure concentration tested.

Munley (1996) exposed pregnant New Zealand rabbits (20 per group) to HFC-236fa by inhalation at daily (6 hr per day) exposure concentrations of 0, 5,000, 20,000, or 50,000 ppm on days 7 to 19 of gestation. Does were killed on day 29, and fetuses were weighed and examined for external, internal, and skeletal abnormalities. There was no evidence of any maternal or developmental toxicity at any exposure concentration tested. There were no compound-related effects on maternal body weight, weight change, or food consumption or in clinical observations or post-mortem findings. There were no compound-related developmental effects. The end points evaluated were mean fetal weight, mean litter size, pre- and post-implantation embryo lethality, and fetal malformations and variations.

Genotoxicity

Bentley (1995a) evaluated HFC-236fa (concentrations ranging from 200,000 to 1,000,000 ppm) for clastogenic activity in human lymphocytes in vitro following 3-hr exposures with and without metabolic activations (S9). No increases in the percent of chromosomally abnormal cells occurred at any HFC-236fa concentration evaluated, and no concentration-related trends in chromosomal-aberration induction were observed.

An inhalation micronucleus study was conducted in male and female mice exposed to HFC-236fa at 0, 5,000, 20,000, or 50,000 ppm for 6 hr per day for 2 consecutive days. Bone-marrow smears were prepared approximately 24 and 48 hr after the second exposure (Bentley 1995b). No statistically significant increases in micronucleated polychromatic erythrocytes were observed in any animal at any concentration tested.

HFC-236fa was also evaluated for mutagenicity in *Salmonella typhimurium* strains TA100, TA1535, TA97, and TA98 and in *Escherichia coli* WPSuvrA- (pKM101) with and without an exogenous metabolic activation system (S9) (Bentley 1995c). At concentrations tested between 0 and 1,000,000 ppm, no evidence of mutagenic activity was detected in either of two independent trials.

On the basis of the available data, the subcommittee concludes that HFC-236fa is not genotoxic and is unlikely to induce heritable effects in humans

Carcinogenicity

No chronic carcinogenicity exposure studies of HFC-236fa are currently available.

SUMMARY

Studies on the metabolism and disposition of HFC-236fa indicate that HFC-236fa is not metabolized to any significant extent. PB-PK modeling using data from gas-uptake studies did not adequately describe the rapid disappearance of HFC-236fa from exposure chambers. Inhalation uptake of HFC-236fa by rats was biphasic with a rapid equilibration phase of up to 30 min followed by a slow linear uptake phase. The pattern of loss was inconsistent with either first-order or saturable metabolism. Experiments have revealed that HFC-236fa has low blood/tissue partition coefficients.

Acute and subchronic inhalation studies in rats have shown HFC-236fa to be of low toxicity. The principal effects of HFC-236fa appear to be attenuation of alerting response, narcosis, and cardiac sensitization. There is some evidence suggesting accommodation to decrements in alerting response resulting from HFC-236fa exposure. Exposure-related alterations in some clinical chemistry measurements have been shown following acute and subchronic exposure of rats to high concentrations of HFC-236fa, but the changes are of questionable biological or toxicological relevance. Reductions in hepatic β-oxidation were observed in rats following 2-week discontinuous exposure to 50,000 ppm, but there were no histopathological correlates to confirm peroxisome proliferation. There was no evidence of peroxisome proliferation in rats following 90-day discontinuous exposure to HFC-236fa at concentrations as high as 50,000 ppm. Studies in dogs indicate that HFC-236fa might have the potential for cardiac sensitization at high exposure concentrations (i.e., ≥150,000 ppm). There have been no studies assessing the carcinogenic potential of HFC-236fa. NOAEL and LOAEL values for noncancer end points are summarized in Table 2-1.

Results of developmental toxicity studies in rats and rabbits suggest that HFC-236fa at concentrations as high as 50,000 ppm is not a developmental toxicant. Signs of maternal toxicity were evident at concentrations of 20,000 and 50,000 ppm in rats but not in rabbits.

Genotoxicity studies of HFC-236fa are negative.

There are no currently available reproductive toxicity or carcinogenicity studies of HFC-236fa.

EXPOSURE GUIDANCE LEVELS

The Navy proposes to use the same exposure guidance levels for HFC-236fa that were set for chlorofluorocarbons CFC-12 and CFC-114 (1-hr EEGL of 2,000 ppm, 24-hr EEGL of 1,000 ppm, and 90-day CEGL of 100 ppm), but did not provide an adequate rationale for doing this. To evaluate the validity the Navy's proposed guidance levels, the subcommittee reviewed the available toxicity data on HFC-236fa to determine what levels would be ade-

TABLE 2-1 Summary of Noncancer Toxicity Information for HFC-236fa

Species	Exposure Frequency and Duration	End Point	NOAEL, ppm	LOAEL, ppm	Reference
Acute Toxicity					
Dog	5 min	Cardiac sensitization	100,000	150,000	HRC 1994
Rat	4 hr	Narcosis	≥134,000	≤189,000	Keller 1994
Rat	4 hr	Narcosis	ND	457,000	Ulrich 1996
Subchronic Toxicity					
Rat	6 hr/d, 5 d	No significant effect	50,000	ND	Smith et al. 1997
Rat	6 hr/d, 5 d/wk for 2 wk	Alerting response decrement	5,000	20,000	Valentine 1995
Rat	6 hr/d, 5 d/wk for 14 wk	Alerting response decrement	20,000	50,000[a]	Valentine 1996
Rat	6 hr/d, gestation d 7-16	Developmental toxicity Maternal toxicity	50,000 5,000	ND 20,000[b]	Munley 1995
Rabbit	6 hr/d, gestation d 7-19	Developmental toxicity Maternal toxicity	50,000 50,000	ND ND	Munley 1996

[a] Accommodation to alerting-response decrement occurred by day 18; exposure-related clinical-chemistry changes were also observed but were not considered to be biologically or toxicologically relevant.
[b] Decreased body-weight gain in dams over first 2 days of exposure; diminished alerting response during exposure.
Abbreviation: ND, not determined.

quately protective of submariner health. A comparison of those results is presented below.

Submarine Exposure Guidance Levels for HFC-236fa

Exposure Level	NRC's Calculated Values	Navy's Proposed Values
1-hr EEGL	10,000 ppm	2,000 ppm
24-hr EEGL	2,000 ppm	1,000 ppm
90-day CEGL	350 ppm	100 ppm

The subcommittee evaluated the Navy's proposed 1-hr EEGL of 2,000 ppm for HFC-236fa by considering an acute toxicity study in rats (Keller 1994) and a cardiac sensitization study in dogs (HRC 1994). In the study by Keller (1994), the NOAEL for a 4-hr exposure was 134,000 ppm based on the end point of narcosis in rats. In the HRC (1994) study, the NOAEL for a 5-min exposure was 100,000 ppm based on cardiac sensitization in dogs. The subcommittee divided those NOAELs by an uncertainty factor of 10 to account for interspecies differences, yielding values of 13,400 ppm based on the Keller (1994) study and 10,000 ppm based on the HRC (1994) study. The Navy's proposed 1-hr EEGL of 2,000 ppm is well below those values, and, therefore, the subcommittee concludes that it is adequately protective of submariner health.

To evaluate the Navy's proposed 24-hr EEGL of 1,000 ppm for HFC-236fa, the subcommittee considered a 2-week toxicity study (Valentine 1995) and a 14-week toxicity study (Valentine 1996) conducted in rats. The 2-week study involved a cumulative exposure of 60 hr and reported a LOAEL of 20,000 ppm and a NOAEL of 5,000 ppm based on alerting response. In contrast, the 14-week study reported a NOAEL of 20,000 ppm based on the same end point. The subcommittee questioned the use of alerting response for determining a NOAEL, because the effects were transient in both studies. In addition, alerting responses are subjectively analyzed and vary among strains and individual animals, and it is unclear how or if such effects are applicable to humans. The subcommittee considered using a NOAEL of 5,000 ppm found in the 2-week study (Valentine 1995) but decided that that value was probably too conservative. Having no data on HFC-236fa at concentrations between 5,000 ppm and 20,000 ppm, the subcommittee believed that 20,000 ppm was a better choice for a NOAEL. The subcommittee divided that value by an uncertainty factor of 10 to account for interspecies differences, yielding a value of 2,000 ppm. The Navy's proposed 24-hr EEGL of 1,000 ppm is half this value and is thus an adequate exposure guidance level.

In its evaluation of the Navy's proposed 90-day CEGL of 100 ppm for HFC-236fa, the subcommittee used the 14-week study in rats (Valentine 1996). On the basis of transient decrements in alerting response, a NOAEL of 20,000 ppm was determined. The subcommittee divided that value by a factor of 10 to account for interspecies differences to yield a value of 2,000 ppm. Because the study involved a discontinuous exposure regimen, the subcommittee multiplied that value by 1/4 (to account for exposure for 6 hr per day) and by 5/7 (to account for exposure five times per week), which yielded a value of about 350 ppm. The Navy's proposed 90-day CEGL of 100 ppm is below that value, and, therefore, the subcommittee concludes that the Navy's CEGL is adequately protective of health for prolonged exposures.

The subcommittee noted that a developmental toxicity study in rats (Munley 1995) reported reduced weight gain and decrements in alerting response in the dams at 20,000 ppm, the same concentration as the NOAEL identified in the 14-week toxicity study. Because the purpose of this project was to establish exposure guidance levels for use on submarines, vessels that have no female crew members, the subcommittee did not use the developmental toxicity study as the basis for calculating the 24-hr EEGL or 90-day CEGL. However, if HFC-236fa is considered for use on vessels with female personnel, those values might have to be reconsidered on the basis of maternal toxicity.

RECOMMENDATIONS

Uncertainties exist with regard to the effects that HFC-236fa might have on human performance. End points of narcosis and decrements in alerting response have been observed in laboratory animals, but it is unclear whether human performance would be affected similarly. Because HFC-236fa was relatively nontoxic in laboratory studies and because human studies have been conducted with other HFCs, such as HFC-23 and HFC-134a, the subcommittee recommends that tests be conducted with humans to determine whether HFC-236fa affects performance skills, such as motor coordination and alertness.

REFERENCES

Bentley, K.S. 1995a. In Vitro Assay of HFC-236fa for Chromosome Aberrations in Human Lymphocytes. Haskell Laboratory Report No. 604-94. Haskell Laboratory, Wilmington, DE.

Bentley, K.S. 1995b. Mouse Bone Marrow Micronucleus Assay of HFC-236fa by Inhalation. Haskell Laboratory. Report No. 602-94. Haskell Laboratory, Wilmington, DE.

Bentley, K.S. 1995c. Mutagenicity Testing of HFC-236fa in the Salmonella Typhimurium and Escherichia Coli Plate Incorporation Assay. Haskell Laboratory Report No. 647-94. Haskell Laboratory, Wilmington, DE.

HRC (Huntington Research Centre). 1994. HFC 236fa. Assessment of Cardiac Sensitization Potential in Dogs. Report No. DPT 293/931308. Huntington Research Centre, Ltd. Huntington, England.

Keller, D.A. 1994. Acute Inhalation Toxicity of HFC-236fa and HFC-236ea in the Rat. Report No. 761-93. Haskell Laboratory for Toxicology and Industrial Medicine. Newark, DE.

Munley, S.M. 1996. Inhalation Developmental Toxicity of HFC-236fa in Rabbits. Haskell Laboratory Report No. 417-96. Haskell Laboratory, Wilmington, DE.

Munley, S.M. 1995. Inhalation Developmental Toxicity Study of HFC-236fa in Rats. Haskell Laboratory. Report No. 66-95. Haskell Laboratory, Wilmington, DE.

Smith, N.D., T.G. Brna, C.L. Gage, and R.V. Hendriks. 1997. New Chemical Alternative for Ozone-Depleting Substances: HFC-236fa. EPA-600/R-97-065. U.S. Environmental Protection Agency, National Risk Management Research Laboratory, Air Pollution Prevention and Control Division, Research Triangle Park, N.C.

Ulrich, C.E. 1996. Acute Inhalation Toxicity Study of HFC-236fa in Albino Rats. WIL Research Laboratories, Inc. WIL-189022. DuPont Report No. HLO 74-96. Ashland, OH.

Valentine, R. 1996. 90-Day Inhalation Toxicity Study with HFC-236fa in Rats. Haskell Laboratory Report No. 211-95; DuPont HLR 211-95. Haskell Laboratory for Toxicology and Industrial Medicine. Newark, DE.

Valentine, R. 1995. Two-week Inhalation Toxicity Study with HFC-236fa in Rats. Haskell Laboratory for Toxicology and Industrial Medicine. Report. No. 596-94; DuPont HLR 596-94, Newark, DE.

Vinegar, A., G.W. Buttler, M.C. Caracci, and J.D. McCafferty. 1995. Gas Uptake Kinetics of 1,1,1,3,3,3-hexafluoropropane (HFC-236fa) and Identification of Its Potential Metabolites. Armstrong Laboratory, Occupational and Environmental Health Directorate, Toxicology Division, Human Systems Center, Air Force Materiel Command. Wright-Patterson A.F.B., OH. AL/OE-TR-1995-0177, NMRI-95-46.

3

Hydrofluorocarbon-23

HYDROFLUOROCARBON (HFC)-23, or trifluoromethane, is a combustion product of HFC-236fa. It belongs to the class of halocarbons. As discussed in Chapter 2, HFC-236fa is under consideration for use in centrifugal chillers aboard naval submarines. If HFC-236fa is accidentally leaked, it will pass through the submarine's carbon-monoxide-hydrogen burner system, which operates at 500°F. Under these conditions, less than 0.1 parts per million (ppm) of HFC-23 is formed per 100 ppm of HFC-236fa. Assuming a worst-case scenario of an HFC-236fa leak at concentrations approaching 100 ppm and no HFC-23 being removed from the air, concentrations of HFC-23 within a submarine could rise by 0.5 ppm per day (Naval Surface Warfare Center 1997).

Emergency exposure guidance levels (EEGLs) and continuous exposure guidance levels (CEGLs) are needed to avoid adverse health effects in submariners from short-term or prolonged exposures to HFC-23 and to avoid degradation in crew performance. This chapter presents the available toxicity information on HFC-23 and the subcommittee's evaluation of the U.S. Navy's proposed 1-hr and 24-hr EEGLs and 90-day CEGL.

CHEMICAL AND PHYSICAL PROPERTIES

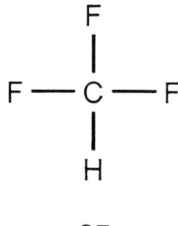

Common name:	HFC-23
Chemical name:	Trifluromethane
Synonyms:	Freon 23, Genetron 23, fluoroform, HC-23, fluoryl, Arcton 1, carbon trifluoride, halocarbon 23, methyl trifluoride, R 23, FE-13
CAS number:	75-46-7
Structural formula:	HCF_3
Description:	Colorless gas
Molecular weight:	70.01
Boiling point:	−82.03°C at 760 mm Hg
Melting point:	−155.2°C
Density and specific gravity:	0.670 g/mL at 25°C
Vapor pressure:	686 psig at 25°C (77°F)
Vapor density:	2.4 (Air = 1.0)
Flash point and flammability:	Nonflammable
Solubility:	0.10 wt% in water
Autoignition:	765°C (1409°F)
Octanol and water partition coefficient:	K_{ow} = 0.64
Conversion factors:	1 mg/m^3 = 0.35; 1ppm = 2.86 mg/m^3

TOXICOKINETICS

Ewing et al. (1990) studied the use of HFC-23 in nuclear magnetic resonance spectroscopy as a gaseous indicator of cerebral blood flow in cats. HFC-23 at a concentration of 600,000 ppm was rapidly absorbed from the lungs and the arterial blood concentration reached a plateau within 3 or 4 min after initiation of the exposure. Likewise, HFC-23 was rapidly eliminated from the blood upon termination of the exposure. It was also rapidly taken up by the brain, with the brain uptake lagging the arterial blood uptake by about 2 min. Similar findings were demonstrated with 670,000 ppm by Detre et al. (1990) in the rat.

No published information is available on the distribution, metabolism, or pharmacokinetics for HFC-23.

TOXICITY INFORMATION

Human Studies

Fagan et al. (1995) conducted a human exposure study to assess the feasibility of using HFC-23 as an indicator for the measurement of human cerebral blood flow in nuclear magnetic resonance spectroscopy. Five subjects were exposed in a blinded fashion to HFC-23 at concentrations ranging from 100,000 to 600,000 ppm. Up to eight pulsed exposures lasting 3 min each, interspersed with 2 min of air exposure between pulses, constituted the exposure regimen. The subjects underwent extensive physiological and neurobehavioral monitoring throughout the study. The maximum tolerated concentration was considered to be 300,000 ppm. Effects such as light-headedness, tingling and numbness of the extremities, and hyperacusis (abnormal acuteness of hearing) were noted at concentrations of 300,000 ppm and above. These effects were seen in only one subject at 300,000 ppm, and no effects were observed at concentrations of 100,000 ppm and 200,000 ppm. No clinically significant physiological effects (e.g., heart rate and rhythm, respiratory rate, and oxygenation) were noted, nor were abnormal clinical laboratory or neurobehavioral effects observed. A small but statistically significant retention of carbon dioxide occurred at 300,000 ppm. The effects of HFC-23 were very short in duration and all subjects were back to baseline at the 30-day evaluation.

Acute Toxicity

HFC-23 has low acute toxicity by the inhalation route. Its approximate lethal concentration (ALC) in rats is greater than 663,000 ppm after a 4-hr inhalation exposure. Kennedy and Nash (1980) exposed groups of six male albino rats to HFC-23 for a single 4-hr period at concentrations of 18,900 ppm, 186,000 ppm, and 663,000 ppm. Oxygen concentrations were maintained at about 20%. Clinical signs observed during exposure to the two higher concentrations included reduced response to sound, gasping, labored breathing, sluggishness, and gnawing. Mild weight loss was observed 24-48 hr after exposure, but normal weight gain was achieved thereafter.

Clayton et al. (1960) reported an ALC of greater than 200,000 ppm in guinea pigs after a 2-hr exposure to HFC-23. Twelve male albino guinea pigs were exposed at that concentration for 2 hr, and no clinical signs or pathological changes were attributable to HFC-23.

Fleming (1945) exposed two male albino guinea pigs to HFC-23 at a concentration of approximately 30,000 ppm for 6 hr. No effects on respiration or weight gain were observed, and no gross or microscopic pathological effects were found when the animals were sacrificed 1 week after exposure.

Cardiac Sensitization

On the basis of studies in three species that are discussed in this section, HFC-23 has an extremely low cardiac-sensitization potential. These studies were done using a standard epinephrine challenge test similar to that reported by Reinhardt et al. (1971). In all the cardiac-sensitization studies discussed below, supplemental oxygen was given at HFC-23 concentrations of 500,000 ppm and above.

Hopkins and Krantz (1968) exposed female mongrel dogs to HFC-23 at 800,000 ppm and found no effect on cardiac automaticity after an intravenous injection of epinephrine. A control injection of epinephrine at 10 μg/kg body weight was given before exposure to HFC-23, and a challenge injection (same dose) was given after a 5- to 10-min exposure period. An electrocardiogram (ECG) was recorded at the beginning of the injection and for at least 60 sec after each epinephrine injection. A formula for measuring the intensity of a myocardial-sensitization response in terms of relative incidence of multifocal ventricular ectopic contractions (RIMVEC) was used to quantitate the electrocardiographic response. The average RIMVEC response following the challenge injection of epinephrine after exposure to HFC-23 was essentially the same as that following the control injection of epinephrine in the five dogs tested.

Hardy and Kieran (1993) exposed male beagle dogs to HFC-23 at concentrations ranging from 100,000 to 500,000 ppm and found no evidence of cardiac sensitization after an intravenous injection of epinephrine. A control injection of epinephrine was given before exposure, and a challenge injection (same dose) was given after 5 min of exposure; the exposure continued for an additional 5 min. The dose of epinephrine was calibrated for each dog to produce an acceptable number (about 10) of ectopic ventricular beats. A dose of 12 μg/kg was the maximum dose used. One dog was tested at each of the six concentrations used. No positive responses were observed in any of the dogs.

Branch et al. (1994) evaluated the acute cardiac and central-nervous-system effects of HFC-23 in eight anesthetized baboons. An intravenous injection of 1 $\mu g/kg$ epinephrine was given after 3 min of exposure to HFC-23 at concentrations of 600,000 ppm or 700,000 ppm. The effects of a control injection of epinephrine had been previously evaluated. Premature ventricular activity totaling four beats was observed in one animal 1 min after the epinephrine injection during inhalation of HFC-23 at 700,000 ppm. No other arrhythmic activity was observed during this study.

Ewing et al. (1990) exposed anesthetized cats to HFC-23 at a concentration of 700,000 ppm for 10 min and observed cardiac sensitization in three of five cats after an intravenous injection of epinephrine. Two of the three sensitized cats received 1 $\mu g/kg$ and the other one received 10 $\mu g/kg$. Two other cats that given epinephrine at 0.1 $\mu g/kg$ did not develop any cardiac arrhythmias.

Branch et al. (1990) exposed anesthetized cats to HFC-23 at a concentration of 700,000 ppm for 10 min and observed cardiac sensitization in three of seven cats after an intravenous injection of epinephrine. All seven cats received 1 $\mu g/kg$. Two other cats that received epinephrine at 0.1 $\mu g/kg$ did not develop any cardiac arrhythmias. In both of these studies, the epinephrine was given immediately after exposure.

Subchronic Toxicity

Leuschner et al. (1983) exposed groups of 20 male and 20 female rats to HFC-23 at 10,000 ppm for 6 hr per day for 90 consecutive days. Likewise, they exposed groups of three male and three female beagle dogs to HFC-23 at 5,000 ppm for 6 hr per day for 90 consecutive days. Clinical signs, food and water consumption, changes in body weight, clinical chemistry, gross and microscopic pathology, and ECGs on the dogs were evaluated. No abnormalities attributable to the test compound were observed and no premature deaths occurred. The investigators concluded that HFC-23 did not produce any adverse effects in the rats or dogs under the test conditions used.

Reproductive Toxicity

No reproductive toxicity studies of HFC-23 are currently available.

Developmental Toxicity

Munley (1997) exposed groups of 25 mated female rats to HFC-23 at concentrations of 5,000 ppm, 20,000 ppm, or 50,000 ppm on days 7 to 21 of gestation for 6 hr per day. No evidence of any maternal or developmental toxicity was observed at any exposure concentration tested. No compound-related effects were observed on maternal body weights, weight changes, food consumption, clinical observations, or post-mortem findings. No compound-related developmental effects were observed. The end points evaluated were mean fetal weight, mean litter size, measures of pre- and post-implantation embryo lethality, and the incidences of fetal malformations and variations. Thus, the no-observed-adverse-effect level (NOAEL) is 50,000 ppm for developmental and maternal effects. Therefore, the results of this study indicate that HFC-23 is not toxic to the rat conceptus.

Genotoxicity

Foltz and Fuerst (1974) exposed *Drosophila melanogaster* to a presumably high, but unknown, concentration of HFC-23 for 10 min to assess its mutagenic potential. HFC-23 was found to significantly increase mutation rates in *Drosophila* progeny over the control rates. However, the investigators state that an undetermined part of the observed mutagenic effects of the gas might be due to anoxia and warrants further investigation. Lee et al. (1983) in a report of the U.S. Environmental Protection Agency gene-tox program reviewed a large number of studies on the sex-linked recessive lethal (SLRL) test for mutagenesis in *D. melanogaster*, including the study by Foltz and Fuerst (1974) cited above. The report categorized HFC-23 as a compound that could not be classified as positive or negative for mutagenic activity in the *Drosophila* SLRL test because of inadequate sample size.

Andrews (1996a) conducted an Ames *Salmonella typhimurium* reverse mutation assay on HFC-23 by exposing five strains of the bacteria to concentrations ranging from 100,000 to 1,000,000 ppm for 24 hr. The results indicate a negative mutagenic response in all five strains both in the presence and in the absence of metabolic activation.

Longstaff et al. (1984) also conducted an Ames *Salmonella* reverse mutation assay on HFC-23 at concentrations up to 500,000 ppm and obtained negative results in four strains of the bacteria, with and without metabolic activation.

Tice (1996a) conducted a mouse bone-marrow micronucleus assay on HFC-23 by exposing groups of five B6C3F$_1$ mice per sex to concentrations ranging from 130,000 to 500,000 ppm for 6 hr per day for three consecutive days. That exposure was followed by a single sample time 24 hr after the final exposure. Positive and negative controls were included in the study. The results indicate that HFC-23 did not significantly increase the frequency of micronucleated polychromatic erythrocytes (PCEs) in the bone marrow of male or female mice or significantly affect the percentage of PCEs in either sex.

Andrews (1996b) evaluated the ability of HFC-23 to induce gene mutations in the guanine phosphoribosyl-transferase (*gpt*) locus of cultured AS52 Chinese hamster ovary (CHO) cells in the presence or absence of metabolic activation. The cells were exposed to HFC-23 in concentrations ranging from 500,000 to 1,000,000 ppm for 5 hr. Positive and negative controls were included in the study. The results indicate that HFC-23, in either the presence or the absence of metabolic activation, did not induce a significant increase in the mutant frequency at the *gpt* locus in cultured AS52 cells.

Tice (1996b) evaluated the potential of HFC-23 to induce structural chromosomal damage in cultured CHO cells in both the absence and the presence of metabolic activation. The cells were exposed to HFC-23 in concentrations ranging from 500,000 to 1,000,000 ppm for 4 hr. Positive and negative controls were included in the study. HFC-23 induced a significant increase in clastogenic damage at concentrations in air ranging from 800,000 to 1,000,000 ppm in the absence of metabolic activation. Under those same conditions, 100% nitrogen induced a significant increase in chromosomal damage to the same magnitude, suggesting the possibility that the response might reflect changes in oxygen concentrations rather than HFC-23 specifically. In the presence of metabolic activation, HFC-23 and 100% nitrogen induced an increase in chromosomal damage that was not statistically significant.

On the basis of the available data, the subcommittee concludes that HFC-23 is not genotoxic and is unlikely to induce heritable effects in humans.

Carcinogenicity

No chronic exposure or carcinogenicity studies of HFC-23 are currently available.

SUMMARY

Toxicokinetic studies indicate that HFC-23 is readily absorbed by the lungs and that blood concentrations reach a plateau within minutes. HFC-23 is rapidly eliminated from the blood upon termination of exposure.

Table 3-1 summarizes the studies on HFC-23 for cardiac sensitization and Table 3-2 summarizes all the noncancer toxicity studies on HFC-23. The animal toxicity information and human exposure data indicate that HFC-23 has low toxicity. The only significant observations were reduced response to sound in rats exposed at 186,000 ppm and lightheadedness, tingling and numbness of the extremities, and hyperacusis in humans after exposure at 300,000 ppm and above.

Cardiac-sensitization studies conducted in three species indicate that HFC-23 has the potential to cause cardiac sensitization at high concentrations. Evidence of such sensitization was observed in baboons and cats exposed at 700,000 ppm, but no evidence was observed in dogs exposed at concentrations as high as 800,000 ppm.

A developmental toxicity study in rats exposed to HFC-23 at a concentration of 50,000 ppm reported no evidence of maternal or fetal toxicity or teratogenicity.

There is no evidence that HFC-23 is genotoxic.

There are no available long-term toxicity studies or carcinogenicity bioassays of HFC-23.

TABLE 3-1 Summary of Cardiac Sensitization Studies with HFC-23

Species	Concentration, ppm	Ventricular Arrhythmia	Reference
Mongrel dogs	800,000	0/5	Hopkins and Krantz 1968
Beagle dogs	100,000-500,000	0/6	Hardy and Kieran 1993
Baboons	600,000	0/8	Branch et al. 1994
Baboons	700,000	1/8	Branch et al. 1994
Cats	700,000	3/5	Ewing et al. 1990
Cats	700,000	3/7	Branch et al. 1990

TABLE 3-2 Summary of Noncancer Toxicity Information for HFC-23

Species	Exposure Frequency and Duration	End Point	NOAEL, ppm	LOAEL, ppm	Reference
Acute Toxicity					
Human	3 min, air 2 min, 8 pulses	Central nervous system effects	200,000	300,000	Fagan et al. 1995
Rat	4 hr	Central nervous system effects	18,900	186,000	Kennedy and Nash 1980
Guinea pig	2 hr	No significant effect	200,000	ND	Clayton et al. 1960
Guinea pig	6 hr	No significant effect	30,000	ND	Fleming 1945
Dog	5-10 min	Cardiac sensitization	800,000	ND	Hopkins and Krantz 1968
Dog	10 min	Cardiac sensitization	500,000	ND	Hardy and Kieran 1993
Baboon	3 min	Cardiac sensitization	600,000	700,000	Branch et al. 1994
Cat	10 min	Cardiac sensitization	ND	700,000	Ewing et al. 1990
Cat	10 min	Cardiac sensitization	ND	700,000	Branch et al. 1990
Subchronic Toxicity					
Rat	6 hr/d, 90 consecutive d	No significant effect	10,000	ND	Leuschner et al. 1983
Dog	6 hr/d, 90 consecutive d	No significant effect	5,000	ND	Leuschner et al. 1983
Developmental Toxicity					
Rat	6 hr/d, gestation d 7-21	Maternal toxicity	50,000	ND	Munley 1997
		Fetal toxicity	50,000	ND	

Abbreviation: ND, not determined.

EXPOSURE GUIDANCE LEVELS

The Navy proposes to use the same exposure guidance levels for HFC-23 that were set for chlorofluorocarbons CFC-12 and CFC-114 (1-hr EEGL of 2,000 ppm, 24-hr EEGL of 1,000 ppm, and 90-day CEGL of 100 ppm), but did not provide an adequate rationale for doing this. To evaluate the validity of the proposed guidance levels, the subcommittee reviewed the available toxicity data on HFC-23 to determine what levels would be adequately protective of submariner health. A comparison of those results is presented below. Because the submariner population is all male, young, and healthier than the general population, the subcommittee did not use an uncertainty factor to account for intraspecies differences in its calculations.

Submarine Exposure Guidance Levels for HFC-23

Exposure Level	NRC's Calculated Levels	Navy's Proposed Levels
1-hr EEGL	20,000 ppm	2,000 ppm
24-hr EEGL	5,000 ppm	1,000 ppm
90-day CEGL	500 ppm	100 ppm

To evaluate the proposed 1-hr EEGL of 2,000 ppm for HFC-23, the subcommittee considered the human exposure study by Fagan et al. (1995). In that study, eight exposures of 3 min each, interspersed with 2 min of air, did not produce any effects at a concentration of 200,000 ppm, and only one of five subjects was affected at 300,000 ppm. Although the cumulative exposure in the study was 24 min, the subcommittee used the NOAEL of 200,000 ppm without time extrapolation because absorption of hydrofluorocarbons via the inhalation route is rapid, reaching maximal concentrations in the blood within 5 min of exposure and equilibrium within the next 15 min (Azar et al. 1973; Trochimowicz et al. 1974; Mullin et al. 1979). However, it was necessary to account for the uncertainty associated with the discontinuous exposure by dividing the NOAEL by an uncertainty factor of 10 to yield a value of 20,000 ppm. Because that value is 10-fold higher than the 1-hr EEGL of 2,000 ppm proposed by the Navy, the subcommittee concludes that the Navy's value is adequately protective of health.

In its evaluation of the Navy's proposed 24-hr EEGL, the subcommittee used a developmental toxicity study in rats (Munley 1997). Although developmental toxicity is not necessarily the most appropriate end point for deriving an EEGL for use aboard submarines (vessels with no female crew members) this study has the most relevant exposure duration (total of 90 hr) of all the available studies, and no maternal or developmental toxicity was observed. Thus, the highest concentration tested of 50,000 ppm was considered to be the NOAEL. Because the available human data on HFC-23 are inadequate to determine the magnitude of difference between rats and humans, the subcommittee divided the NOAEL by an uncertainty factor of 10 to account for interspecies variability to yield a value of 5,000 ppm. The Navy's proposed 24-hr EEGL of 1,000 ppm is five times lower than that value and is therefore an adequate guidance level.

To evaluate the Navy's proposed 90-day CEGL, the subcommittee used the 90-day toxicity studies in rats and dogs (Leuschner et al. 1983). For rats, the NOAEL was 10,000 ppm for 6 hr per day for 90 consecutive days, and for dogs, the NOAEL was 5,000 ppm for the same duration. The NOAELs in these studies were the only concentrations tested. The subcommittee believes that they are probably lower than the true NOAEL for HFC-23 because NOAELs reported in 90-day studies of similar HFCs were higher; for example, the NOAEL was 40,000 ppm for HFC-143a (Brock et al. 1996) and 50,000 ppm for HFC-125 (Nakayama et al. 1993 as cited in Kawano et al. 1995) and HFC-134a (Hext 1989; Collins et al. 1995). Furthermore, HFC-23 had no maternal or fetal effects in a developmental toxicity study at a concentration of 50,000 ppm. Given this, an uncertainty factor to adjust for the 6 hr per day exposures was not used. However, an uncertainty factor of 10 was applied to the NOAELs to account for interspecies variability, giving values of 1,000 ppm and 500 ppm, respectively. Because the Navy's proposed 90-day CEGL of 100 ppm is 5-fold lower than 500 ppm, the subcommittee finds the Navy's exposure guidance levels to be adequately protective of health for prolonged exposures.

REFERENCES

Andrews, P.W. 1996a. Salmonella Typhimuriam Microsome Reverse Mutation Assay. Proj. No. ILS A073-001. Integrated Laboratory Systems, Durham, N.C.

Andrews, P.W. 1996b. AS52/GPT Mammalian Mutagenesis Assay. Proj. No. ILS A073-003. Integrated Laboratory Systems, Durham, N.C.

Azar, A., H.J. Trochimowicz, J.B. Terrill, and L.S. Mullin. 1973. Blood levels of fluorocarbon related to cardiac sensitization. Am. Ind. Hyg. Assoc. J. 34:102-109.

Branch, C.A., D.A. Goldberg, J.R. Ewing, S.C. Fagan, S.S. Butt, and J. Gayner. 1994. Evaluation of the acute cardiac and central nervous system effects of the fluoro-

carbon trifluoromethane in baboons. J. Toxicol. Environ. Health 43:25-35.
Branch, C.A., J.R. Ewing, S.C. Fagan, D.A. Goldberg, and K.M.A. Welch. 1990. Acute toxicity of a nuclear magnetic resonance cerebral blood flow indicator in cats. Stroke 21:1172-1177.
Brock, W.J., H.J. Trochimowicz, C.H. Farr, R.J. Millischer, and G.M. Rusch. 1996. Acute, subchronic, and developmental toxicity and genotoxicity of 1,1,1-trifluoroethane (HFC-143a). Fundam. Appl. Toxicol. 31:200-209.
Clayton, J.W., Jr., D.B. Hood, and J.W. Williams. 1960. Acute Testing. Rep. No. 25-60. Haskell Laboratory, Newark, DE.
Collins, M.A., G.M. Rusch, F. Sato, P.M. Hext, and R.J. Millischer. 1995. 1,1,1,2-Tetrafluoroethane: repeat exposure inhalation toxicity in the rat, developmental toxicity in the rabbit, and genotoxicity in vitro and in vivo. Fundam Appl. Toxicol. 25:271-280.
Detre, J.A., C.J. Eskey, and A.P Koretsky. 1990. Measurement of cerebral blood flow in rat brain by 19F-NMR detection of trifluoromethane washout. Magn. Reson. Med. 15:45-57.
Ewing, J.R., C.A. Branch, S.C. Fagan, J.A. Helpern, R.T. Simkins, S.M. Butt, and K.M.A. Welch. 1990. Fluorocarbon-23 measure of cat cerebral blood flow by nuclear magnetic resonance. Stroke 21:100-106.
Fagan, S.C., A.A. Rahill, G. Balakrishnan, J.R. Ewing, C.A. Branch, and G.G. Brown. 1995. Neurobehavioral and physiologic effects of trifluoromethane in humans. J. Toxicol. Environ. Health 45:221-229.
Fleming, A.J. 1945. Kitchen Tests on "Freon" Refrigerants. Rep. No. 0023-45. Haskell Laboratory, Newark, DE.
Foltz, V.C., and R. Fuerst. 1974. Mutation studies with Drosophila melanogaster exposed to four fluorinated hydrocarbon gases. Environ. Res. 7:275-285.
Hardy, C.J., and P.C Kieran. 1993. Halon 13B1, Freon 23, Mixture of Freon 23 and HFC 125, Assessment of Cardiac Sensitisation Potential in Dogs. DPT 273/921009. Huntingdon Research Centre Ltd., Huntingdon, Cambridgeshire, England.
Hext, P.M. 1989. HFC-134a: 90-Day Inhalation Toxicity Study in the Rat. ICI Rep. No. CTL/P/2466. Central Toxicology Laboratory, Imperial Chemical Industries, Alderley Park, Macclesfield, Cheshire, U.K.
Hopkins, R.M., and J.C. Krantz. 1968. Relative effects of haloforms and epinephrine on cardiac automaticity. Anesth. Analg. 47:56-67.
Kennedy, G.L., and S.D. Nash. 1980. Inhalation Approximate Lethal Concentration. Rep. No. 641-80. Haskell Laboratory, Newark, DE.
Lee, W.R., S. Abrahamson, R. Valencia, E.S. von Halle, F.E. Wurgler, and S. Zimmering. 1983. The Sex-linked recessive lethal test for mutagenesis in Drosophila melanogaster, a report of the U.S. Environmental Protection Agency Gene-Tox Program. Mutation Res. 123:183-279.
Leuschner, F., B.W. Neumann, and F. Hubscher. 1983. Report on subacute toxicological studies with several fluorocarbons in rats and dogs by inhalation. Arzneim.-Forsch. 33:1475-1476.

Longstaff, E., M. Robinson, C. Bradbrock, J.A. Styles, and I.F.H. Purchase. 1984. Genotoxicity and carcinogenicity of fluorocarbons: assessment by short-term in vitro tests and chronic exposure in rats. Toxicol. Appl. Pharmacol. 72:15-31.

Mullin, L.S., C.F. Reinhardt, and R.E. Hemingway. 1979. Cardiac arrhythmias and blood levels associated with inhalation of Halon 1301. Am. Ind. Hyg. Assoc. J. 40:653-658.

Munley, S.M. 1997. HFC-23: Inhalation Developmental Toxicity Study in Rats. Rep. No. 995-96. Haskell Laboratory, Newark, DE.

Naval Surface Warfare Center. 1997. Subject: Refrigerant decomposition in submarine $CO-H_2$ burners. Letter to Commander, Naval Sea Systems Command, from Commander, Carderock Division, Naval Surface Warfare Center, Philadelphia, PA., dated June 24, 1997.

Reinhardt, C.F., A. Azar, M.E. Maxfield, P.E. Smith, Jr., and L.S. Mullin. 1971. Cardiac arrhythmias and aerosol "sniffing." Arch. Environ. Health 22:265-279.

Tice, R.R. 1996a. Repeated Inhalation Exposure of FE-13 in Mice, Mus musculus (Bone Marrow Micronucleus Assay). Proj. No. ILS A073-002. Integrated Laboratory Systems, Durham, N.C.

Tice, R.R. 1996b. In Vitro Chromosome Aberrations Study in Chinese Hamster Ovary (CHO) Cells. Proj. No. ILS A073-004. Integrated Laboratory Systems, Durham, N.C.

Trochimowicz, H.J., A. Azar, J.B. Terrill, and L.S. Mullin. 1974. Blood levels of fluorocarbon related to cardiac sensitization: Part II. Am. Ind. Hyg. Assoc. J. 35:632-639.

4

Hydrofluorocarbon-404a

HYDROFLUOROCARBON (HFC)-404a is a gaseous mixture of three halocarbons—52% HFC-143a, 44% HFC-125, and 4% HFC-134a. It is used as a refrigerant in ice-cream machines. The U.S. Navy is considering installing a single ice-cream machine aboard each of its submarines. The Navy does not intend to perform servicing of this equipment when its submarines are under way, so the only HFC-404a aboard will be in the ice cream machine (i.e, there will be no cylinders of HFC-404a aboard). The refrigerant systems of the machines are sealed and operate in a manner similar to that of a household refrigerator.

To protect submariners from large accidental releases or low-level slow releases of HFC-404a, emergency exposure guidance levels (EEGLs) and continuous exposure guidance levels (CEGLs) are needed to avoid adverse health effects from short-term or prolonged exposures to HFC-404a and to avoid degradation in crew performance. No toxicity studies on HFC-404a were available, so the subcommittee reviewed the available data on its three components—HFC-143a, HFC-125, and HFC-134a. The subcommittee used these data to determine 1-hr and 24-hr EEGLs and 90-day CEGLs for each of the components, and then those values were used to calculate exposure guidance levels for HFC-404a on the basis of percent composition of the individual components. The calculated EEGLs and CEGL for HFC-404a were used to evaluate the Navy's proposed exposure guidance levels for HFC-404a.

HFC-143a

Chemical and Physical Properties

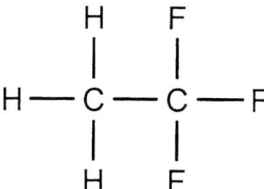

Common name:	HFC-143a
Chemical name:	1,1,1-Trifluoroethane
Synonyms:	FC-143a; hydrofluorocarbon 143a
CAS number:	420-46-6
Structural formula:	CH_3CF_3
Description:	Colorless gas
Molecular weight:	84.04
Boiling point:	−47.3°C at 760 mm Hg
Melting point:	−111.3°C
Flash point and flammability:	−90°C
Solubility:	Soluble in chloroform and ether
Auto-ignition:	750°C
Conversion factors:	1 ppm = 3.40 mg/m^3; 1 mg/m^3 = 0.29 ppm

Toxicokinetics

In a study of rats exposed by inhalation to HFC-143a at 4,800 ppm for 4-5 hr, 2,2,2-trifluoroethanol was the only metabolite detected. Nuclear magnetic resonance showed that HFC-143a was metabolized slowly, primarily through an oxidative route (DuPont 1994, as cited in AIHA 1996).

Toxicity Information

Acute Toxicity

Groups of six male rats were exposed for 4 hr (nose only) to HFC-143a at concentrations of 97,000 or 540,000 parts per million (ppm) (Brock et al. 1996). No deaths occurred during exposure or during the 14-day post-exposure observation period. No clinical signs attributed to HFC-143a were observed. Slight-to-moderate body-weight losses were observed in both exposure groups on the day following exposure, but normal weight gains were observed through the remainder of the study. Thus, the 4-hr LC_{50} (concentration causing death in 50% of test animals) for rats is considered to be greater than 540,000 ppm.

The LC_{50} value of HFC-143a in mice is reported to be greater than 500,000 ppm (Nikijenko and Tolgskaya 1965, as cited in AIHA 1996).

Cardiac Sensitization

Brock et al. (1996) evaluated the cardiac-sensitization potential of HFC-143a in an epinephrine challenge test. An intravenous control injection of epinephrine (2-12 μg/kg) was administered to groups of six beagle dogs that were subsequently exposed for 10 min to HFC-143a at concentrations ranging from 50,000 to 300,000 ppm via single-pass-through face mask. Five minutes after initiating vapor exposure, a challenge concentration of epinephrine (same as the pre-test concentration) was administered. Evidence of cardiac sensitization was determined if multiple ectopic beats (more than five beats) or ventricular fibrillation, which could be fatal, was evident. No cardiac-sensitization responses were observed at HFC-143a concentrations of 50,000 to 250,000 ppm. At 300,000 ppm, two of five dogs were considered to have exhibited cardiac sensitization.

Subchronic Toxicity

Four groups of rats (10 of each sex) were exposed nose-only for 6 hr per day, 5 days per week for 4 weeks to HFC-143a at 0, 2,000, 10,000, or 40,000 ppm (Brock et al. 1996). Clinical signs, body weights, food consumption, clinical pathology, organ weights, and tissue histopathology were evaluated. Premature deaths in three rats from different exposure groups were considered incidental. The only remarkable finding was degenerative changes in the testes of male rats in all exposure groups. No evidence of toxicity was observed in female rats. Excessive temperature conditions and problems with the proper fit of the nose-only restraint system on the animals were reported. To investigate whether the testicular changes might have been an artifact of the excessive temperature conditions, which are known to adversely affect the testes of rats and humans (Van Demark and Fre 1970), or the stress of the restraint system, another 4-week study was conducted in male rats using the same HFC-143a concentrations under normal chamber temperature conditions and without restraint (i.e., whole-body exposure system) (Brock et al. 1996). No adverse testicular effects or adverse clinical signs were observed at any exposure concentration in the study. Because of the absence of testicular changes in this study as well as in a 90-day study (discussed below), the subcommittee believes that the testicular changes observed in the first 4-week study were not caused by exposure to HFC-143a.

Four groups of rats (20 of each sex) were exposed (whole body) for 6 hr per day, 5 days per week for 90 days to 0, 2,000, 10,000, or 40,000 ppm (Brock et al. 1996). Clinical signs, body weights, food consumption, clinical pathology, organ weights, and tissue histopathology were evaluated. Liver β-oxidation activity, an indicator of peroxisomal proliferation, was also measured. There were no HFC-143a-related adverse effects at any exposure level.

Reproductive Toxicity

No reproductive toxicity studies of HFC-143a are currently available.

Developmental Toxicity

Two inhalation studies, one in rats and one in rabbits, were conducted to evaluate the developmental toxicity of HFC-143a.

Brock et al. (1996) exposed groups of 25 pregnant rats to HFC-143a at concentrations of 0, 2,000, 10,000, or 40,000 ppm for 6 hr per day on days 6 to 15 of gestation. There were no signs of maternal toxicity during or after exposure, nor were there any significant changes in body weight, body-weight gains, or food consumption throughout the study. The number of corpora lutea and implants and the incidence of malformations and developmental variations in the fetuses were not statistically significant. However, there was a significant increase in the mean percentage per litter of visceral variations due to retarded development in all the test groups. When these anomalies were combined for visceral, skeletal, and external variations, a statistically significant increase was found only at the highest concentration. However, the increase appeared to be a reflection of an unusually low incidence (1.6%) of variations in the control group (the range in historical controls was 6.8% to 16.2%). Because of that, the investigators did not consider the observed anomalies to be compound related. No external variations were observed in any group.

In a study with rabbits, groups of 24 pregnant animals were exposed to HFC-143a at concentrations of 0, 2,000, 10,000, or 40,000 ppm for 6 hr per day on days 6 to 18 of gestation (Brock et al. 1996). There were no signs of maternal toxicity during or after exposure, nor were there any significant changes in body weight, body-weight gains, or food consumption throughout the study. However, there was a slight increase in the incidence of combined malformations, expressed as mean percentage per litter, in the litters of the 2,000-ppm and 40,000-ppm groups. The increase appeared to be primarily due to increases in skeletal malformations (expressed as percentage incidence per litter); increases of 1.5%, 7.5%, 3.4%, and 6.3% were found in the control, 2,000-ppm, 10,000-ppm, and 40,000-ppm groups, respectively. Because no clear concentration response was found, however, in either the types or the numbers of malformations and the incidence was within the range of historical controls (0-12.9%), the investigators did not consider the malformations to be compound related. No statistically significant increase in the incidence of soft tissue or external malformation was found compared with controls.

Genotoxicity

Ames *Salmonella* reverse mutation assays were conducted in three independent laboratories using a modified experimental design for testing HFC-143a gas (Longstaff et al. 1984; Brock et al. 1996). In one laboratory, HFC-143a was mutagenic in two of four *Salmonella* strains tested (TA1535 and TA100). HFC-143a was not mutagenic in tests performed at the other two laboratories that used four to five *Salmonella* strains, including TA1535 and TA100. It was also negative in an *Escherichia coli* strain (WP2uvrA) assay system which detects DNA damage and repair.

HFC-143a was negative in a cell-transformation (Styles) assay using BHK21 cultures (Longstaff et al. 1984). HFC-143a was not clastogenic in in vitro tests with cultured human lymphocytes at exposure concentrations up to 35,000 ppm (Brock et al. 1996). There was no statistically significant increase in micronuclei in bone-marrow cells of male and female mice exposed to concentrations of up to 40,000 ppm for 6 hr per day for 2 consecutive days (Brock et al. 1996).

On the basis of the available data, the subcommittee concludes that HFC-143a is not genotoxic and is unlikely to induce heritable effects in humans.

Carcinogenicity

HFC-143a was one of five fluorocarbons tested for carcinogenicity in rats (36 of each sex) using a limited bioassay design (Longstaff et al. 1984). Each fluorocarbon was dissolved in corn oil and a dose of 300 mg/kg was administered by gavage for 5 days per week for 52 weeks. Control groups consisted of an undosed group (32 per sex) and two vehicle dosed groups (36-40 per sex). Clinical signs, body weights, gross abnormalities, and tissue (lungs, liver, spleen, kidneys, and brain) histopathology were evaluated. The study was terminated at week 125. Male rats receiving HFC-143a had lower mean body weights from weeks 28 to 88. Mortality in the exposed group was similar to that in the control groups. There was no significant increase in incidence of neoplasms in any organ in the HFC-143a exposure group.

Exposure Guidance Levels for HFC-143a

A summary of the noncancer toxicity studies on HFC-143a is presented in Table 4-1. On the basis of those data, the subcommittee calculated 1-hr and 24-hr EEGLs and a 90-day CEGL for HFC-143a. Because the submariner population is all male, young, and healthier than the general population, the subcommittee did not use an uncertainty factor to account for intraspecies differences in its calculations.

For a 1-hr EEGL, a cardiac-sensitization study in dogs (Brock et al. 1996) was found to be the most appropriate for determining a NOAEL of 250,000 ppm. Because absorption of hydrofluorocarbons via the inhalation route is rapid, reaching maximal concentrations in the blood within 5 min of exposure and equilibrium within the next 15 min (Azar et al. 1973; Trochimowicz et al. 1974; Mullin et al. 1979), the NOAEL identified for cardiac sensitization following a 10-min exposure can be used without time extrapolation.

TABLE 4-1 Summary of Noncancer Toxicity Information for HFC-143a

Species	Exposure Frequency and Duration	End Point	NOAEL, ppm	LOAEL, ppm	Reference
Acute Toxicity					
Rat	4 hr	No significant effect	540,000	ND	Brock et al. 1996
Dogs	10 min	Cardiac sensitization	250,000	300,000	Brock et al. 1996
Subchronic Toxicity					
Rat	6 hr/d, 5 d/wk for 4 wk	Testicular changes	ND	2,000[a]	Brock et al. 1996
Rat	6 hr/d, 5 d/wk for 4 wk	No significant effect	40,000	ND	Brock et al. 1996
Rat	6 hr/d, 5 d/wk for 90 d	No significant effect	40,000	ND	Brock et al. 1996
Developmental Toxicity					
Rat	6 hr/d, gestation d 6-15	Maternal toxicity Fetal toxicity	40,000 40,000	ND ND	Brock et al. 1996
Rabbit	6 hr/d, gestation d 6-18	Maternal toxicity Fetal toxicity	40,000 40,000	ND ND	Brock et al. 1996

[a]End point considered to be an artifact of exposure system (nose-only exposure and excessive temperature conditions), because repeated study under normal exposure (whole body) and temperature conditions did not cause similar effects.
Abbreviation: ND, not determined.

The subcommittee divided the NOAEL by an uncertainty factor of 10 to account for interspecies variability, because there are no human data on HFC-143a, for a 1-hr EEGL of 25,000 ppm.

For determining a 24-hr EEGL, the subcommittee used a 4-week toxicity study in rats, in which the highest tested concentration of 40,000 ppm was the NOAEL (Brock et al. 1996). The NOAEL was divided by an uncertainty factor of 10 to account for interspecies variability for a 24-hr EEGL for HFC-143a of 4,000 ppm.

A 90-day toxicity study (Brock et al. 1996) in rats was used to calculate the 90-day CEGL for HFC-143a. In that study, the highest tested concentration of 40,000 ppm was the NOAEL. The subcommittee divided that value by a factor of 10 to account for interspecies variability and then multiplied that value by 1/4 (to account for exposure for 6 hr per day) and by 5/7 (to account for exposure 5 days per week), which yielded a value of about 700 ppm.

HFC-125

Chemical and Physical Properties

Common name:	1,1,1,2,2-Pentafluoroethane
Chemical name:	1,1,1,2,2-Pentafluoroethane
Synonyms:	Pentafluoroethane; HFC-125; fluorocarbon 125
CAS number:	354-33-6
Structural formula:	CF_3CHF_2
Description:	Colorless gas
Molecular weight:	120.0
Boiling point:	−48.5°C
Density and specific gravity:	1.35 g/cc at 21°C
Vapor pressure:	1,381 psia at 25°C
Vapor density:	4 (air = 1)
Flash point and flammability:	Nonflammable
Solubility:	0.97 g/L in water at 25°C
Autoignition:	No applicable

Octanol and water partition
coefficient: $\log P_{ow} = 1.48$
Conversion factors: 1 ppm = 4.90 mg/m^3;
1 mg/m^3 = 0.20 ppm

Toxicokinetics

In a study of male rats exposed to HFC-125 at 10,000 ppm for 6 hr, Harris et al. (1992) demonstrated that HFC-125 is slowly metabolized to trifluoroacetic acid. The rate of metabolism of HFC-125 was shown to be much slower than that of hydrochlorofluorocarbon (HCFC)-124 (2-chloro-1,1,1,2-tetrafluoroethane) or HCFC-123 (2,2-dichloro-1,1,1-trifluoroethane). That finding is consistent with preliminary data (PAFT 1989), as reported by ECETOC (1994), showing that HFC-125 undergoes little uptake and metabolism at exposure concentrations ranging from 1,000 to 50,000 ppm.

Another study (Nakayama et al. 1993, as cited in Kawano et al. 1995) reported no detectable increases in plasma or urine fluoride concentrations in rats exposed to HFC-125 at concentrations up to 50,000 ppm for 4 or 13 weeks, which also suggests that metabolism of HFC-125 is low.

Wang et al. (1993) reported that HFC-125 stimulates oxygen consumption and the defluorination of 2-chloro-1,1-difluoroethane in hepatic microsomes from phenobarbital treated rabbits. As in the other studies, no metabolism of HFC-125 was detected under the incubation conditions used.

Toxicity Information

Summaries of the toxicology of HFC-125 have been published by the European Centre for Ecotoxicology and Toxicology of Chemicals (ECETOC 1994) and Kawano et al. (1995).

Acute Toxicity

Groups of five male rats were exposed for 4 hr to HFC-125 at concentrations of 504,000 or 710,000 ppm (Panepinto 1990, as cited in ECETOC 1994). There was no mortality or clinical signs, but transient body-weight loss was observed. In another study, rats (five of each sex) were exposed (whole body) to HFC-125 at 800,000 ppm for 4 hr (Nakayama et al. 1992a, as cited in Kawano et al. 1995). Oxygen was maintained at 200,000 ppm. No deaths occurred during exposure or during the 14-day post-exposure observation period. Clinical signs observed during exposure included ataxia, decreased locomotor activity, dyspnea, and decreased auditory response. These signs disappeared within 1 hr after exposure. Thus, the 4-hr LC_{50} value for HFC-125 is considered to be greater than 800,000 ppm.

Cardiac Sensitization

Male beagle dogs were exposed to HFC-125 at concentrations of 0, 75,000, 100,000, 150,000, 200,000, 250,000, or 300,000 ppm for 10 min. An intravenous injection of epinephrine was administered before and during exposure to HFC-125 (Hardy 1992, as cited in Kawano et al. 1995). If a life-threatening arrhythmia occurred after the challenge injection, the material was considered a cardiac sensitizer at that inhaled concentration. A known cardiac sensitizer (CFC-11) was used as a positive control to validate the system, and Halon 1301 (CF_3 Br), also a known cardiac sensitizer, was used to provide comparative data. Positive evidence of cardiac sensitization was seen at 100,000 ppm and greater but not at 75,000 ppm. Therefore, the no-observed-adverse-effect level (NOAEL) for cardiac sensitization was determined to be 75,000 ppm and the lowest-observed-adverse-effect level (LOAEL) was 100,000 ppm. HFC-125 was less potent than CFC-11 but more potent than Halon 1301.

Vinegar and Jepson (1995) proposed a quantitative approach for relating blood-concentration time courses to cardiac-sensitization thresholds during inhalation of HFC-125. A physiologically based pharmacokinetic model was used to simulate blood concentrations of HFC-125 in humans during inhalation exposure. The target concentration of HFC-125 in blood was established by simulating a 5-min inhalation exposure at the LOAEL for cardiac sensitization (100,000 ppm). Although the chemical concentration in venous blood does not achieve a steady-state value after 5 min, that exposure period is used in most cardiac-sensitization protocols prior to epinephrine challenge. The blood concentration achieved after 5 min was used as the concentration at which cardiac sensitization was likely to occur. The exposure time required to produce that target level at various concentrations was estimated for resting- and moderate-activity conditions. Results of this study showed that some exposures will not produce the target chemical concentrations in blood no matter how long the exposure occurs, that exposure concentrations well above the NOAEL can reach the target chemical concentrations in blood within min, and that exposure concentrations below the LOAEL can reach target chemical concentrations in blood if exposure continues long enough.

Subchronic Toxicity

Nakayama et al. (1992b, as cited in Kawano et al. 1995) exposed (whole body) rats (10 of each sex per group) to HFC-125 for 6 hr per day, 5 days per week for 4 weeks at concentrations of 0, 5,000, 15,600, or 50,000 ppm. In addition, 10 rats per sex were assigned to the control and high-exposure groups for assessment of a 2-week post-exposure recovery. Clinical observations, body weight, food consumption, clinical pathology, organ weight, and tissue histopathology were evaluated. Liver β-oxidation activity was also measured. There were no HFC-125-related adverse effects at any exposure concentration.

In another study using the same test concentrations as above (5,000, 15,600, and 50,000 ppm), groups of rats (20 per sex) were exposed for 13 weeks, and the same biological end points measured in the 4-week study were evaluated (Nakayama et al. 1993, as cited in Kawano et al. 1995). No adverse effects were observed.

Reproductive Toxicity

No reproductive toxicity studies of HFC-125 are currently available.

Developmental Toxicity

Two inhalation studies, one in rats and one in rabbits, were conducted to evaluate the developmental toxicity of HFC-125.

In the study with rats (Master et al. 1992, as cited in Kawano et al. 1995), groups of 25-29 pregnant rats were exposed (whole body) to HFC-125 at concentrations of 0, 5,000, 15,000, or 50,000 ppm for 6 hr per day on days 6-15 of gestation. The only signs of clinical maternal effects were observed in the group exposed at 50,000 ppm. The rats exhibited unsteady gait during exposure but not after. There were no significant differences between the exposed groups and the control group in any of the litter indices evaluated (i.e., number of corpora lutea and implants, pre-implant loss, embryonic deaths, post-implant loss, number of live young, litter weights, and mean fetal weights) or in the incidences of fetal malformations, visceral abnormalities, or skeletal anomalies.

In the study with rabbits, groups of 16 pregnant rabbits were exposed (whole body) to HFC-125 at concentrations of 0, 5,000, 15,000, and 50,000 ppm for 6 hr per day on days 6-18 of gestation (Brooker et al. 1992, as cited in Kawano et al. 1995). The adult rabbits exposed to HFC-125 at 50,000 ppm exhibited transient lower weight gain and food consumption but no other adverse clinical signs. The litters were evaluated for the same litter indices evaluated in the rat study and for malformations and anomalies. No significant changes in those end points were found in any of the exposed groups compared with controls.

Genotoxicity

Ames bacterial reverse mutation assays were conducted in two independent laboratories using a modified experimental design for testing HFC-125 gas (Longstaff et al. 1984; May et al. 1992, as cited in ECETOC 1994). At concentrations ranging from 200,000 to 1,000,000 ppm, HFC-125 was not mutagenic in five *Salmonella* strains (TA98, TA100, TA1535, TA1537, and TA1538). It was also negative in an *Escherichia coli* strain (WP2uvrA) assay, which detects DNA damage and repair.

At concentrations ranging from 175,000 to 700,000 ppm, HFC-125 was not clastogenic in an in vitro cytogenic assay using human lymphocytes (Dance et al. 1992, as cited in ECETOC 1994). HFC-125 was also evaluated for chromosomal aberration using Chinese hamster ovary cells at concentrations up to 600,000 ppm (Dance and Hodson-Walker 1992, as cited in ECETOC 1994). There was no clear evidence of clastogenic activity at any of the test concentrations after 4 or 24 hr of exposure, but after 48 hr, an increase in both frequency of aberrant cells and incidence of polyploid cells was observed at 600,000 ppm.

There was no statistically significant increase in micronuclei in the bone-marrow erythrocytes of male and female mice exposed at concentrations of up to 600,000 ppm for 6 hr (Edwards et al. 1992, as cited in ECETOC 1994). Mice exposed to HFC-125 at a concentration of 600,000 ppm exhibited transient tremors, hunched posture, and weight loss at 24 hr post-exposure.

Exposure Guidance Levels for HFC-125

A summary of the noncancer toxicity studies on HFC-125 is presented in Table 4-2. On the basis of those data, the subcommittee calculated 1-hr and 24-hr EEGLs and a 90-day CEGL for HFC-125. Because the submariner population is all male, young, and healthier than the general population, the subcommittee did not use an uncertainty factor to account for intraspecies variability in its calculations.

To calculate a 1-hr EEGL, the subcommittee used a cardiac sensitization study in dogs, in which the NOAEL was 75,000 ppm (Hardy 1992, as cited in Kawano et al. 1995). Because absorption of hydrofluorocarbons via the inhalation route is rapid, reaching maximal concentrations in the blood within 5 min of exposure and equilibrium within the next 15 min (Azar et al. 1973; Trochimowicz et al. 1974; Mullin et al. 1979), the NOAEL for the 10-min exposure can be used without time extrapolation. The NOAEL was divided by an uncertainty factor of 10 to account for interspecies differences, as no human data are available on HFC-125, to yield a 1-hr EEGL of 7,500 ppm.

For the 24-hr EEGL, the subcommittee used a 4-week toxicity study in rats, in which the highest tested concentration of 50,000 ppm was the NOAEL (Nakayama et al. 1992b, as cited in Kawano et al. 1995). The NOAEL was divided by an uncertainty factor of 10 to extrapolate from animals to humans to yield a 24-hr EEGL of 5,000 ppm.

In a 90-day toxicity study in rats, the highest tested concentration of 50,000 ppm was the NOAEL for the study (Nakayama et al. 1993, as cited in Kawano et al. 1995). That value was divided by an uncertainty factor of 10 to account for interspecies variability, and that value was then multiplied by 1/4 (to account for exposure for 6 hr per day) and by 5/7 (to account for exposure five times per week), which yielded a 90-day CEGL of about 900 ppm.

To calculate the 24-hr EEGL and the 90-day CEGL for HFC-125, the subcommittee used a NOAEL of 50,000 ppm. However, one developmental toxicity study (Master et al. 1992, as cited in Kawano et al. 1995) reported that pregnant rats exhibited an unsteady gait at a concentration of 50,000 ppm. The subcommittee did not use this sign of toxicity for determining the NOAEL, because the exposure guidance levels being developed for the Navy are specifically for use on submarines—vessels with no female crew members. If HFC-125 (or HFC-404a) is used aboard other vessels, the exposure guidance levels should be re-evaluated on the basis of maternal toxicity.

A summary of the toxicology of HFC-134a was published by ECETOC (1995). In addition, the National Research Council published an evaluation

TABLE 4-2 Summary of Noncancer Toxicity Information for HFC-125

Species	Exposure Frequency and Duration	End Point	NOAEL, ppm	LOAEL, ppm	Reference
Acute Toxicity					
Rat	4 hr	No significant effect	710,000	ND	Panepinto 1990 (as cited in ECETOC 1994)
Rat	4 hr	Ataxia, decreased locomotor activity, dyspnea, and decreased auditory response	ND	800,000	Nakayama et al. 1992a (as cited in Kawano et al. 1995)
Dog	10 min	Cardiac sensitization	75,000	100,000	Hardy 1992 (as cited in Kawano et al. 1995)
Subchronic Toxicity					
Rat	6 hr/d, 5 d/wk for 4 wk	No significant effect	50,000	ND	Nakayama et al. 1992b (as cited in Kawano et al. 1995)
Rat	6 hr/d, 5 d/wk for 13 wk	No significant effect	50,000	ND	Nakayama et al. 1993 (as cited in Kawano et al. 1995)
Developmental Toxicity					
Rat	6 hr/d, gestation d 6-15	Maternal toxicity Fetal toxicity	15,000 50,000	50,000 ND	Master et al. 1992 (as cited in Kawano et al. 1995)
Rabbit	6 hr/d, gestation d 6-18	Maternal toxicity Fetal toxicity	50,000 50,000	ND ND	Brooker et al. 1992 as cited in Kawano et al. 1995

Abbreviation: ND, not determined.

of the toxicity of HFC-134a in a report titled *Toxicity of Alternatives to Chlorofluorocarbons: HFC-134a and HCFC-123* (NRC 1996). Below is a brief description of the toxicity data presented in that evaluation and in new studies on HFC-134a.

HFC-134a

Chemical and Physical Properties

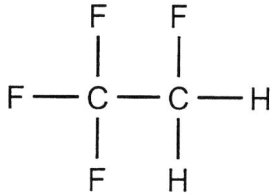

Common name:	FC-134
Chemical name:	1,1,1,2-Tetrafluoroethane
Synonyms:	HFC-134a; norflurane, 1,1,1,2-tetrafluoroethane; HFA-134a; 1,2,2,2-tetrafluoroethane; F-134a; R134a; refrigerant R134a
CAS number:	811-97-2
Structural formula:	CF_3CFH_2
Description:	Colorless gas
Molecular weight:	102.03
Boiling point:	$-26.5°C$ at 736 mm Hg
Melting point:	$-101°C$
Density and specific gravity:	1.21 g/mL (liquid under pressure at 77°F)
Vapor pressure:	96 psia at 25°C
Flash point and flammability:	Nonflammable
Solubility:	0.15% in water; soluble in ether
Octanol and water partition coefficient:	$P_{pw} = 1.06$
Conversion factors:	1 ppm = 4.20 mg/m³
	1 mg/m³ = 0.24

Toxicokinetics

Finch et al. (1995a) studied the absorption of HFC-134a by exposing (head only) male and female rats to 150,000 ppm for 1 hr and using ^{19}F nuclear magnetic resonance (NMR) imaging. The study showed that HFC-134a reached a steady state after 25 min of exposure, at which 58.3 ± 11.9 mg had been absorbed per rat (100-225 g). HFC-134a was rapidly eliminated after exposure; the elimination half-life was 4.6 ± 0.6 min in male rats and 4.9 ± 1.5 min in female rats. In another study with rats, the mean half life of HFC-134a was 7.8 ± 1.5 min after a 10-sec exposure and 8.1 ± 1.7 min after a 10-min exposure (Finch et al. 1995b).

Ellis et al. (1993) exposed male and female rats to [U-^{14}C]HFC-134a at a concentration of 10,000 ppm for 1 hr. Of the inhaled dose, approximately 1% was recovered in the recovered in the urine, feces, and expired air, indicating that absorption of HFC-134a was poor. Approximately two-thirds of the recovered [U-^{14}C]HFC-134a were exhaled within 1 h after exposure as unchanged HFC-134a, and
the remainder was detected in carbon dioxide or excreted in the urine or feces as trifluoroacetic acid. About 0.15% of the inhaled dose of [U-^{14}C]HFC-134a was retained in the body 5 days after exposure. The highest concentrations were found in the lung followed by the heart and then plasma, liver, and spleen. Total metabolism was about 0.37% of the inhaled dose. Trifluoroacetic acid was the only fluorinated metabolite detected in the urine by ^{19}F NMR spectroscopy.

In vitro studies with hepatocytes from rats and liver microsomes from humans, rats, and rabbits indicate that HFC-134a is metabolized by cytochrome P-450 (Olson et al. 1990a,b; Olson and Surbrook 1991). The maximal rate of HFC-134a metabolism by human microsomes was low, showed little interindividual variation among the samples evaluated, and did not exceed that observed with rat or rabbit microsomes. The cytochrome P-450 isoform CYP 2E1 is the major isoform that catalyzes the biotransformation of HFC-134a in rat and human liver microsomes, but isoforms CYP 2B1 (phenobarbitol inducible), CYP 1A2 (β-naphthoflavone inducible), and CYP 1A1 also catalyze the biotransformation of HFC-134a (Olson et al. 1991; Surbrook and Olson 1992).

On the basis of the in vitro and in vivo metabolic data, the fate of HFC-134a can be described by the scheme shown in Figure 4-1.

Monte et al. (1994) studied the metabolism of HFC-134a in humans using ^{19}F NMR spectroscopy. Four male volunteers were administered HFC-134a as a single 1,200-mg dose using a metered-dose inhaler. Trifluoroacetic acid

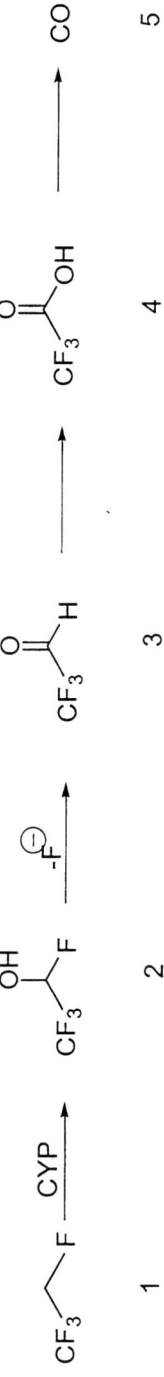

FIGURE 4-1 Metabolic fate of HFC-134a. 1, HFC-134a; 2, 1,3,3,3-tetrafluoroethanol; 3, trifluoroacetaldehyde; 4, trifluoroacetic acid; 5, carbon dioxide.

was the only fluorinated metabolite detected in the urine. The total amount of trifluoroacetic acid detected ranged from none to 5,186 ng (0.00039% of the inhaled dose), which indicates that oxidative metabolism of HFC-134a occurs in humans but accounts for only a small fraction of the administered dose.

In another study (Pike et al. 1995), seven subjects were exposed to HFC-134a at a dose of 75 mg using a metered-dose inhaler, and the absorption, distribution, and retention of HFC-134a was determined using a whole-body γ-counter. HFC-134a was rapidly eliminated in exhaled air within minutes after exposure. An average of 9.6% of the inhaled dose was retained in the body after 5 min. That remaining dose was eliminated with an apparent terminal half-life of 1.5-4.2 hr; on average, less than 1% of the administered dose was retained in the body at 5.8 hr. Approximately 0.0056% of the administered dose of HFC-134a was found in the urine within the first 2 hr, and later samples contained no significant amounts. HFC-134a was distributed to all regions of the body and then cleared without evident accumulation in a specific region.

As part of an ascending concentration safety study of HFC-134a (described later in this chapter), Emmen and Hoogendijk (1999) collected pharmacokinetic data from human subjects. Four men and four women were exposed (whole body) to HFC-134a at concentrations of 1,000, 2,000, 4,000, and 8,000 ppm for 1 hr. A summary of the pharmacokinetic data are presented in Table 4-3. The mean maximum blood concentration of HFC-134a ranged from 1.02 to 7.22 μg/mL. Blood concentrations increased in an exposure-related but nonlinear pattern. Blood concentrations of HFC-134a were higher in males than in females. The total amount of HFC-134a absorbed (expressed as the area under the curves extrapolated to infinity, $AUC_{0-\infty}$) also suggests an exposure-related pattern, with higher values for males than for females. These findings are supported by measurements taken during exposure, after exposure, and for both periods combined. The distribution phase of the elimination half-life for HFC-134a was approximately 9 min for females and 10 min for males, and the terminal phase of the elimination half-life was about 43 min for females and 42 min for males. No exposure-related patterns were observed for these values.

Toxicity Information

Human Studies

HFC-134a is being developed for use in pharmaceutical applications. The International Pharmaceutical Aerosol Consortium for Toxicity Testing

TABLE 4-3 Human Pharmacokinetic Data on HFC-134a from the Emmen and Hoogendijk (1999) Study[a]

Exposure Concentration, ppm	C_{max} (μg/mL)	$T_{1/2}$ (min) (D)	$T_{1/2}$ (min) (T)	$AUC_{0-\infty}$ (μg*min/mL)	AUD_1 (μg*min/mL)	AUD_2 (μg*min/mL)	AUD_T (μg*min/mL)
Females							
1,000	1.02 (0.17)	5.70[b] (0.76)	39.10 (19.74)	74.14 (7.88)	49.37 (5.91)	20.51 (6.98)	69.88 (10.27)
2,000	1.44 (0.38)	9.74[c] (1.83)	57.57 (63.56)	110.46 (47.57)	62.94 (16.91)	27.33 (4.74)	90.28 (20.79)
4,000	3.06 (0.18)	11.08[b] (3.89)	31.78 (10.14)	209.35 (9.30)	135.52 (12.98)	57.84 (14.88)	193.36 (5.49)
8,000	5.95 (0.53)	9.44[c] (3.88)	44.45 (34.14)	462.21 (42.74)	280.20 (21.12)	125.54 (8.96)	405.75 (16.50)
Males							
1,000	1.02 (0.18)	9.81 (4.25)	37.95 (12.72)	73.64 (10.82)	48.72 (9.22)	17.58 (5.30)	66.30 (9.65)
2,000	1.92 (0.40)	9.00[c] (1.73)	34.34 (12.72)	141.96 (28.00)	92.86 (22.87)	38.94 (9.62)	131.80 (27.04)
4,000	3.79 (0.47)	11.21 (4.23)	57.76 (34.70)	304.93 (81.29)	173.96 (29.05)	73.37 (22.91)	247.33 (24.10)
8,000	7.22 (0.70)	8.34 (1.18)	38.29 (2.33)	534.22 (82.43)	342.87 (45.02)	139.66 (24.94)	482.53 (67.98)

[a]Values are mean (SD), n = 4
[b]Value is mean of n = 3. No $T_{1/2}$(D) could be calculated for one subject.
[c]Value is mean of n = 2. No $T_{1/2}$(D) could be calculated for two subjects.

Abbrieviations: n, number of subjects; SD, standard deviation; C_{max}, highest concentration; $T_{1/2}$(D), distribution half-life (elimination phase) (α); $T_{1/2}$(T), terminal half-life (elimination phase) (β); $AUC_{0-\infty}$, area under the curve - extrapolation from last measured data point; AUD_1, area under the data - during exposure; AUD_2, area under the data - after exposure; AUD_T, area under the data - during and after exposure.

(IPACTI) is conducting a development program to satisfy regulatory requirements for the use of HFC-134a in drug applications. Harrison et al. (1996) evaluated the safety and tolerability of HFC-134a in a 28-day double-blind inhalation aerosol study using healthy, nonsmoking males between 18 and 55 years of age. Subjects used a metered-dose inhaler and received either four inhalations (puffs) four times per day for 14 days or eight inhalations four times a day for 14 days. The subjects were then crossed over to the alternative exposure regimen with the same propellant for the next 14-day period. The metered-dose inhaler contained HFC-134a as a propellant, ethanol as a co-solvent, and oleic acid as a surfactant. Subjects were instructed to hold their breath for 10 sec following each inhalation and to wait 30 sec before taking the next puff. There were no clinically significant changes in blood pressure, heart rate, electrocardiograms (ECGs), pulmonary function, hematology, or serum chemistry. One of eight subjects in the HFC-134a group had elevated eosinophil counts throughout the study. There were no serious or unexpected adverse events due to inhalation of HFC-134a. Headache was the most frequent adverse event reported by subjects in the double-blind study groups.

Vinegar et al. (1997) exposed human volunteers to HFC-134a via inhalation for the purpose of collecting pharmacokinetic data to validate a human physiologically based pharmacokinetic model. The intended exposure was 4,000 ppm for 30 min. A nonrebreathing valve system was used to deliver HFC-134a, and a catheter was inserted in the antecubital vein of subjects to collect blood samples. Subjects were monitored for ECG, blood pressure, and pulse rate throughout the exposure. Exposures were terminated for safety reasons following unexpected and uncontrollable changes in pulse rate during the inhalation exposure. Due to the small sample size and experimental design, no conclusion or speculation about cause and effect was offered.

Subsequent to the interim report by Vinegar et al. (1997), a consortium of producers and pharmaceutical industry users initiated a human inhalation exposure study with HFC-134a (Emmen and Hoogendijk 1999). Four men and four women were exposed (whole body) to four concentrations of HFC-134a ranging from 1,000 to 8,000 ppm for 1 hr periods. Subjects were assigned to a randomized ascending concentration exposure regimen that included two air-only exposures and two CFC-12 (dichlorodifluoromethane) exposures. Exposures were administered once per week for 8 consecutive weeks. ECG, pulse rate, blood pressure (diastolic and systolic), and pulmonary function (peak expiratory flow) were evaluated. Blood concentrations of HFC-134a and CFC-12 were measured before, during, and after each exposure, and a pharmacokinetic analysis was performed. No exposure-related effects of inhalation exposure to HFC-134a on ECG, pulse rate, blood pressure, and lung function were observed compared with air control and CFC-12 reference conditions. The blood concentrations appeared to approach a plateau between 15 and 30 min into the exposure, although some increases were frequently seen at 55 min. Elimination from the body was rapid.

Acute Toxicity and Irritation

In 1996, the NRC found HFC-134a to have low acute toxicity via the inhalation route. The approximate lethal concentration of HFC-134a for a 4-hr exposure in rats was 567,000 ppm (Silber and Kennedy 1979a), and lethality was observed in two of four rats exposed to 750,000 ppm for 30 min (Rissolo and Zapp 1967). In a study published more recently, the 4-hr LC_{50} value of HFC-134a in rats was reported to be greater than 500,000 ppm (Collins 1984, as cited in ECETOC 1995).

Additional acute inhalation toxicity studies have been published. Alexander and Libretto (1995) conducted acute 1-hr inhalation toxicity tests of HFC-134a in mice, rats, and beagle dogs. No deaths or clinical signs were observed in mice and rats exposed (nose only) to 810,000 ppm (with oxygen supplementation). Dogs exposed (face only) to 160,000 or 320,000 ppm exhibited salivation, head shaking, and struggling, but no mortality. Clinical signs were not observed in dogs exposed to concentrations of 80,000 ppm or lower.

Local irritation and sensitization tests were performed with HFC-134a as either a gas or a liquid. HFC-134a gas was slightly irritating to the eyes of rabbits (Mercier 1990a, as cited in ECETOC 1995), and liquefied HFC-134a was slightly irritating to rabbit skin (Mercier 1989, as cited in ECETOC 1995), although the effect on skin might have been due to local freezing. Liquefied HFC-134a did not cause contact sensitization in guinea pigs (Mercier 1990b, as cited in ECETOC 1995).

Cardiac Sensitization

In 1996, the NRC found HFC-134a to be a weak cardiac sensitizer when tested in two epinephrine challenge tests in dogs (Mullin and Hartgrove 1979; Hardy et al. 1991). On the basis of the Mullin and Hartgrove (1979) study, the NOAEL and LOAEL for HFC-134a were 50,000 ppm and 75,000 ppm, respectively. The Hardy et al. (1991) study indicated a NOAEL of 40,000 ppm and a LOAEL of 80,000 ppm.

The EC_{50} value for cardiac sensitization for HFC-134a was found to be 270,000 ppm (ICI Chemicals and Polymers 1991, as cited in AIHA 1991).

Subchronic Toxicity

Three subchronic inhalation toxicity studies (Silber and Kennedy 1979b; Riley et al. 1979; Hext 1989) were considered in the NRC's earlier evaluation of HFC-134a (NRC 1996). Male rats exposed to HFC-134a at a concentration of 100,000 ppm for 6 hr per day, 5 days per week for 2 weeks had increased respiratory rate and increased urine fluoride levels, but there were no alterations in body weight, organ weight, clinical pathology, or tissue histopathology (Silber and Kennedy 1979b). Riley et al. (1979) exposed rats (16 of each sex per group) to HFC-134a at 0, 1,000, 10,000, or 50,000 ppm for 6 hr per day, 5 days per week for 4 weeks. Minor changes were observed in liver, kidney, and testes weights at 50,000 ppm and 10,000 ppm (liver only), but there were no histopathological changes in these organs. A slight focal interstitial pneumonia was observed in the male rats of the 50,000-ppm group. Clinical observations, body weights, and clinical pathology were normal. The NRC also reviewed an unpublished inhalation study by Hext (1989), which was subsequently published by Collins et al. (1995). In this study, groups of rats (20 males and 20 females) were exposed to HFC-134a at 0, 2,000, 10,000, or 50,000 ppm for 6 hr per day, 5 days per week for 13 weeks. Biological end points evaluated were clinical observations, body weights, food consumption, organ weights, clinical pathology, and tissue histopathology. No exposure-related effects were observed.

Additional subchronic inhalation studies were also evaluated by the subcommittee. In a 90-day study with mice exposed (nose only) to HFC-134a at concentrations up to 50,000 ppm for 1 hr per day, no clinical signs of toxicity and no alteration in body-weight gains were observed (Alexander and Libretto 1995). Similarly, no clinical signs or effects on weight gains were observed in rats exposed daily for 1 hr to HFC-134a at 2,500, 15,000, or 50,000 ppm for 50 weeks (Alexander and Libretto 1995).

Alexander and Libretto (1995) also conducted two separate 1-year inhalation toxicity studies of HFC-134a in dogs using different delivery systems. In one study, dogs received 120,000 ppm for 1 hr per day via a face-mask inhalation delivery system. In the other study, pulse concentrations of up to 2.25 grams were administered twice a day with a metered-concentration inhaler via an oropharyngeal tube. Except for signs of anxiety that were attributed to the delivery systems, there were no exposure-related clinical signs or effects on body weights, food consumption, eye, heart function, respiratory rate, pulse rate, clinical pathology, organ weights, or tissue histopathology.

Reproductive and Developmental Toxicity

In 1996, the NRC evaluated several inhalation toxicity studies (Hodge et al. 1979a; Lu and Staples 1981; Wickramaratne 1989a,b) conducted in animals to examine the reproductive and developmental effects of HFC-134a. In studies of pregnant rats exposed to HFC-134a during days 6-15 of gestation, NOAELs for maternal toxicity were 30,000 ppm (Lu and Staples 1981) and 50,000 ppm (Hodge et al. 1979a). A LOAEL of 100,000 ppm was identified for maternal toxicity, which included reduced responses to noise stimuli and uncoordinated movements (Lu and Staples 1981). With respect to fetal toxicity in rats, significant reductions in fetal body weight and slightly delayed skeletal ossification were observed at 50,000 ppm in one study (Hodge et al. 1979a); in another study, significant reductions in fetal body weight and significant increases in several skeletal variations were observed at 300,000 ppm but not at 100,000 ppm. In a study with rabbits exposed to HFC-134a during gestation, there was no evidence of maternal toxicity at 2,500 ppm, but statistically significant reductions in food consumption and body-weight gain were observed at 10,000 ppm (Wickramaratne 1989a,b). No fetal toxicity was observed in rabbits exposed to HFC-134a at 10,000 ppm.

Additional reproductive and developmental toxicity studies have been published. In a study by Collins et al. (1995), groups of 28 pregnant rabbits were exposed to HFC-134a at concentrations of 0, 2,500, 10,000, or 40,000 ppm for 6 hr per day on days 7-19 of pregnancy. The rate of body-weight gain in the groups exposed to HFC-134a at 10,000 ppm and 40,000 ppm was lower than that in the controls, but the reduction was partially compensated for in the 40,000-ppm group by an increase in body weights after exposure was stopped. The investigators concluded that the difference between maternal body-weight gain in rabbits exposed to HFC-134a at 40,000 ppm and in controls was small and represented slight maternal toxicity and that the very small reduction in body-weight gain observed at 10,000 ppm represented a minimal maternal effect. There were no adverse effects on reproduction, and there was no evidence of fetal toxicity or teratogencity.

Alexander et al. (1996) conducted a fertility study with rats. Groups of 30 male rats and 30 female rats were exposed daily by snout only for 1 hr to HFC-134a at concentrations of 0, 2,500, 10,000, or 50,000 ppm throughout gametogenesis (10 and 3 weeks before pairing for males and females, respectively) and mating. In this parental generation (F_0 generation), males continued to be exposed until week 18 and females were exposed until termination. Fourteen pregnant rats from each exposure group were killed on day 20 postcoitum and their uterine contents and ovaries examined. The remaining females were allowed to litter and rear their young (F_1 generation). To allow for parturition, exposure to HFC-134a was stopped after dosing on day 20 of pregnancy and resumed on days 1-21 postpartum. One male and one female rat from each litter (12 litters per group) were raised to maturity and mated, and survival and development were evaluated in the resulting progeny (F_2 generation). One F_2 rat of each sex from each litter (8 litters per group) was raised to sexual maturity.

No clinical signs of toxicity were observed in the F_0, F_1, or F_2 generations. A slight but statistically significant decrease in body-weight gain was observed in F_0 males after 2 weeks of exposure to HFC-134a at 10,000 and 50,000 ppm, and cumulative and overall weight gains were also reduced in males exposed at 50,000 ppm. However, no effect on weight gain was observed in F_0 females. No significant effects on body weight were observed in the F_1 or F_2 generations. With regard to breeding performance, no effects were observed on estrous cycles, mating, precoital times, conception, and gestation length in the F_0 or F_1 generations. No significant changes in the number of live-born pups, sex ratio, survival postpartum, or development (i.e., pinna detachment, upper incisor eruption, eye opening, reflex development, cleavage of the balanopreputial skinfold, and vaginal opening) were observed in the F_0 and F_1 litters.

No exposure-related abnormalities were observed in any of the postmortem examinations of the F_0, F_1, or F_2 generations. In addition, when the uterine contents of pregnant F_0 rats sacrificed before term were examined, the number of corpora lutea, implants, embryonic deaths, live young, sex ratio, litter weights, and fetal body weights did not differ significantly from controls, and the incidence, type, and distribution of visceral or skeletal abnormalities did not increase.

A study of perinatal and postnatal exposure to HFC-134a was also conducted (Alexander et al. 1996). Groups of 41 mated female rats formed the F_0 generation. These rats were exposed daily (snout only) for 1 hr to HFC-134a at concentrations of 0, 1,800, 9,900, or 64,400 ppm on days 17-20 of pregnancy and then on days 1-21 postpartum. One male and one female from the F_1 generation (20 litters per group) were raised to maturity. The F_1 animals were mated and the pregnant females were killed on day 20 of pregnancy for examination of uterine contents and ovaries.

No clinical signs of toxicity were observed in the F_0 or F_1 generations, nor were there any significant effects on body weights. With regard to breeding performance, no effects were observed on estrous cycles, mating, precoital times, conception, and gestation length in the F_0 or F_1 generations. No significant changes in the number of live-born pups, sex ratio, and survival postpartum were observed. There was a statistically significant delay in pinnae detachment, eye opening, and startle response in the F_1 generation from parents exposed to HFC-134a at 64,400 ppm. However, because the delays were small, they were not considered biologically significant. No exposure-related abnormalities were observed in any of the postmortem examinations of the F_0 and F_1 generations. In addition, when the uterine contents of pregnant females sacrificed before term were examined, the number of corpora lutea, implants, embryonic deaths, live young, sex ratio, litter weights, and fetal body weights did not differ significantly from controls, and the incidence, type, and distribution of visceral or skeletal abnormalities did not increase.

Barton et al. (1994) studied the effects of HFC-134a on testicular endocrine function of male rats. Groups of 25 male rats were exposed daily to HFC-134a at concentrations of 10,000, 30,000, or 100,000 ppm for 11 weeks before mating, during mating, and after mating for a total exposure duration of 18 weeks. Nose-only exposure was given for the first 9 weeks and then whole-body exposure for the remainder of the study. At 16 weeks, the basal level of luteinizing hormone (LH) was high in all groups, including the control group. No intergroup differences were observed when these animals were stimulated with luteinizing-hormone releasing hormone. When the testes from exposed animals were incubated with human chorionic gonadotrophin, there was a small increase in testosterone secretion and biosynthesis with a concomitant rise in progesterone secretion in the 100,000-ppm group but not in the lower-concentration groups. There were no changes in androgen biosynthesis, and levels of 17-hydroxyprogesterone, androstenedione, and estradiol were not significantly different from control levels.

Genotoxicity

Numerous in vitro and in vivo genotoxicity tests were performed with

HFC-134a. HFC-134a was tested in four Ames assays for bacterial mutagenesis in *Salmonella* strains TA98, TA100, TA1535, TA1537, and TA1538 and *E. coli* strain WP2uvrA (Brusick 1976, as cited in ECETOC 1995; Longstaff et al. 1984; Callander and Priestley 1990, as cited in ECETOC 1995; Araki 1991, as cited in ECETOC 1995; Collins et al. 1995). HFC-134a was also tested for yeast mutagenesis in *Saccharomyces cerevisiae* (Brusick 1976, as cited in ECETOC 1995); it was nongenotoxic in all studies even when concentrations were as high as 1,000,000 ppm in some assays. At test atmospheres of more than 90%, inhibition of growth was observed (Collins et al. 1995).

No abnormalities were observed in chromosomal-aberration studies using either human lymphocytes (Mackay 1990 cited in ECETOC 1995) or Chinese hamster lung cells (Asakura 1991, as cited in ECETOC 1995) at concentrations up to 1,000,000 ppm. Toxicity was observed in cultured human lymphocytes at atmospheres of more than 75% at the 96-hr sampling time (Collins et al. 1995). HFC-134a was inactive in a cell-transformation assay using BHK21 cultures (Longstaff et al. 1984).

HFC-134a was not clastogenic in an in vivo mouse micronucleus assay, where male and female mice were exposed for 6 hr at concentrations up to 500,000 ppm (Müller and Hofmann 1989, as cited in ECETOC 1995). HFC-134a did not induce unscheduled DNA synthesis in male rats following exposure for 6 hr at concentrations as high as 100,000 ppm (Trueman 1990, as cited in ECETOC 1995). When male rats were exposed at concentrations up to 50,000 ppm for 6 hr in a single exposure or for 5 consecutive days for bone-marrow cytogenetic evaluation (Anderson and Richardson 1979, as cited in ECETOC 1995), there was no statistically significant increase in total chromosomal aberrations. In a dominant lethal assay, male mice were exposed 6 hr per day for 5 consecutive days at concentrations as high as 50,000 ppm (Hodge et al. 1979b). Males were mated with unexposed virgin females at weekly intervals for several weeks. HFC-134a did not affect fertility or cause mutagenic effects.

Carcinogenicity

Several chronic studies have been conducted with HFC-134a. In 1996, the NRC reviewed one oral study (Longstaff et al. 1984) and one inhalation study (Hext and Parr-Dobrzanski 1993). In the study by Longstaff et al. (1984), rats were orally administered HFC-134a in corn oil at a dose of 300 mg/kg of body weight per day for 5 days per week for 52 weeks. There was no significant increase in incidence of neoplasms in any organ in the HFC-

134a-exposed group. However, the significance of these results is uncertain because the study did not involve lifetime exposure.

The NRC also evaluated an inhalation study by Hext and Parr-Dobrzanski (1993), which was subsequently published by Collins et al. (1995). In that study, rats (85 of each sex per group) were exposed (whole body) to HFC-134a at concentrations of 0, 2,500, 10,000, or 50,000 ppm for 6 hr per day, 5 days per week for 104 weeks. The only exposure-related effect of toxicological importance was an increased incidence of Leydig-cell hyperplasia and Leydig-cell adenoma in male rats in the 50,000-ppm group. The tumors were benign and not life threatening. The survival rate was similar in all groups. In 1996, the NRC concluded that Leydig-cell tumors were not applicable to humans and thus were not considered an adverse effect.

Two additional studies have been published. Alexander et al. (1995) exposed (nose-only) mice (60 of each sex per group) to HFC-134a at concentrations of 2,500, 15,000, or 75,000 ppm for 1 hr per day for at least 104 weeks. There were two air-only (control) exposure groups. Clinical observations, behavioral observations (Modified Irwin Screen), body weights, hematology, and microscopic tissue pathology were monitored. There were no exposure-related effects in any of the measurements.

In another study, rats (60 of each sex per group) were exposed (nose only) to HFC-134 at concentrations of 2,500, 10,000, or 50,000 ppm for 1 hr per day for at least 108 weeks (Alexander et al. 1995). HFC-134a vapor was delivered using a metered-dose inhaler. An air-only (control) exposure group consisted of 120 rats of each sex. Clinical observations, behavioral observations (Modified Irwin Screen), body weights, hematology, and microscopic tissue pathology were monitored. A statistically significant increase in subacute and chronic laryngitis occurred in female rats only in the 50,000-ppm group. The severity of laryngitis was slight, and the observation was considered to be of no toxicological significance. There were no exposure-related effects in any of the other measurements.

Exposure Guidance Levels for HFC-134a

In 1996, the NRC reviewed the available toxicity data on HFC-134a and proposed a 1-hr EEGL of 4,000 ppm, a 24-EEGL of 1,000 ppm, and a 90-day CEGL of 900 ppm. The Navy chose to set lower values for the 1-hr EEGL and 90-day CEGL because of the lack of experience with HFCs, but it did adopt the NRC's proposed 24-EEGL.

Since the 1996 review, additional data on HFC-134a have become available. An updated summary of the noncancer toxicity studies on HFC-134a are presented in Table 4-4. The subcommittee used the new studies to re-evaluate the exposure guidance levels proposed by the NRC in 1996 and those currently used by the Navy (see below). Because the submariner population is all male, young, and healthier than the general public, the committee did not use an uncertainty factor for intraspecies variability in its calculations.

Exposure Guidance Levels for HFC-134a

Exposure Level	NRC's Calculated Levels	NRC's (1996) Calculated Levels	Navy's Levels
1-hr EEGL	8,000 ppm	4,000 ppm	2,000 ppm
24-hr EEGL	5,000 ppm	1,000 ppm	1,000 ppm
90-day CEGL	900 ppm	900 ppm	100 ppm

One of the new studies was an ascending-concentration safety study in humans. The subjects were exposed to HFC-134a at concentrations up to 8,000 ppm for 1 hr with no adverse effects (Emmen and Hoogendijk 1999). The subcommittee believes that this study should be used to determine the 1-hr EEGL rather than the cardiac-sensitization study in dogs (Hardy et al. 1991) used by the NRC (1996) in its earlier evaluation of HFC-134a. Although the human subjects were not challenged with epinephrine as in the dog study, the subcommittee notes that the human NOAEL of 8,000 ppm is five-fold lower than the NOAEL of 40,000 ppm for dogs. Thus, the subcommittee believes that a 1-hr EEGL of 8,000 ppm can be justified. That value is four-fold greater than that currently used by the Navy.

The subcommittee considered a 13-week toxicity study in rats (Hext 1989; Collins et al. 1995) to be the most appropriate for deriving the 24-hr EEGL for HFC-134a. In this study, the highest concentration of 50,000 ppm was the NOAEL. Because the available data on HFC-134a were inadequate to determine the magnitude of difference between rats and humans, the NOAEL was divided by an uncertainty factor of 10 to account for interspecies variability, which yielded a 24-hr EEGL of 5,000 ppm. This exposure level is higher than the 1,000-ppm guidance level used by the Navy and proposed by the NRC in 1996. The reason for the difference is that in 1996 the NRC was determining exposure levels for use aboard Navy ships with female crew members and, therefore, based the NOAEL of 10,000 ppm on a devel-

TABLE 4-4 Updated Summary of Noncancer Toxicity Information for HFC-134a

Species	Exposure Frequency and Duration	End Point	NOAEL, ppm	LOAEL, ppm	Reference
Acute Toxicity					
Dog	10 min	Cardiac sensitization	50,000	75,000	Mullin and Hartgrove 1979
Dog	10 min	Cardiac sensitization	40,000	80,000	Hardy et al. 1991
Mice	1 hr	No significant effect	810,000	ND	Alexander and Libretto 1995
Rat	1 hr	No significant effect	810,000	ND	Alexander and Libretto 1995
Rat	4 hr	Lethality	ND	567,000[a]	Silber and Kennedy 1979a
Dog	1 hr	Salivation, head shaking, and struggling	80,000	160,000	Alexander and Libretto 1995
Subchronic Toxicity					
Rat	6 hr/d, 5 d/wk for 14 d	Increased respiratory rate	ND	100,000	Silber and Kennedy 1979b
Rat	6 hr/d, 5 d/wk for 4 wk	Slight focal interstitial pneumonia	10,000	50,000[b]	Riley et al. 1979
Rat	6 hr/d, 5 d/wk for 13 wk	No significant effect	50,000	ND	Hext 1989; Collins et al. 1995
Rat	1 hr/d, 7 d/wk for 50 wk	No significant effect	50,000	ND	Alexander and Libretto 1995

Species	Exposure	Effect			Reference
Mice	1 hr/d, 7 d/wk for 90 d	No significant effect	50,000	ND	Alexander and Libretto 1995
Dogs	1 hr/d, 7 d/wk for 1 yr	No significant effect	120,000	ND	Alexander and Libretto 1995
Dogs	Twice a day using a metered dose inhaler via oropharyngeal tube for 1 yr	No significant effect	2.25 g		Alexander and Libretto 1995; Alexander et al. 1995
Human	1 hr, 1 d/wk for 8 wk	No significant effect	8,000	ND	Emmen and Hoogendijk 1999

Developmental and Reproductive Toxicity

Species	Exposure	Effect			Reference
Rat	6 hr/d, gestation days 6-15	Maternal toxicity Fetal toxicity	30,000 100,000	100,000 300,000	Lu and Staples 1981
Rat	6 hr/d, gestation days 6-15	Maternal toxicity Fetal toxicity	50,000 10,000	ND 50,000	Hodge et al. 1979a
Rabbit	6 hr/d, gestation days 6-18	Maternal toxicity Fetal toxicity	2,500 10,000	10,000 ND	Wickramaratne 1989a,b
Rabbit	6 hr/d, gestation days 7-19	Maternal toxicity Fetal toxicity	2,500 40,000	10,000 ND	Collins et al. 1995
Rat	1 hr/d, 10 wk (F_0 male) or 3 wk (F_0 female) before mating, during mating, and on days 1-21 postpartum for females	Maternal toxicity (F_0) Paternal toxicity (F_0) Fetal toxicity (F_1 and F_2)	50,000 50,000 50,000	ND ND ND	Alexander et al. 1996

TABLE 4-4 *(Continued)*

Species	Exposure Frequency and Duration	End Point	NOAEL, ppm	LOAEL, ppm	Reference
Rat	1 hr/d on gestation days 17-20 and on days 1-21 postpartum (F_0 generation)	Maternal toxicity (F_0)	64,400	ND	Alexander et al. 1996
		Fetal toxicity (F_1 and F_2)	64,400	ND	
Male rat	6 hr/d for 18 wk	Testicular endocrine function[c]	30,000	100,000	Barton et al. 1994
Rat	1 hr/d, 10 wk (F_0 male) or 3 wk (F_0 female) before mating, during mating, and on days 1-21 postpartum for females	Fertility (F_0 and F_1)	50,000	ND	Alexander et al. 1996
	1 hr/d on gestation days 17-20 and on days 1-21 postpartum (F_0 generation)	Fertility (F_0 and F_1)	64,400	ND	

[a]Approximate lethal concentration.
[b]This effect was not observed in subsequent studies.
[c]Increase in testosterone secretion and biosynthesis and a concomitant increase in progesterone secretion when the testis was incubated with human chorionic gonadotrophin.
Abbreviation: ND, not determined.

opmental study in which fetal toxicity was observed (Hodge et al. 1979a). However, fetal toxicity is not as a relevant an end point for setting an exposure level for use on submarines, which have no female crew members.

In 1996, the NRC used a 2-year chronic toxicity study (Hext and Parr-Dobrzanski 1993) in male rats to derive a 90-day CEGL for HFC-134a. At the highest concentration of 50,000 ppm in the study, increases in testicular weight and benign Leydig tumors were reported. However, the increases in testicular weight were not considered to be an adverse effect in itself, and the increase in Leydig tumors is not applicable to humans, because those tumors are related to a peculiar aspect of rodent metabolism. Thus, the NOAEL for the study was 50,000 ppm. That value was divided by an uncertainty factor of 10 to account for interspecies variability. That product was then multiplied by 1/4 (to account for exposure for 6 hr per day) and by 5/7 (to account for exposure five times per week), which yielded a 90-day CEGL of about 900 ppm. The subcommittee agrees with this earlier determination of the CEGL.

SUMMARY

In summary, the toxicity profiles of the three components of HFC-404a indicate that these chemicals are practically nontoxic and, therefore, there is only marginal concern for potential health hazard. Absorption and metabolism studies in laboratory animals show little uptake of all three components and that metabolism is primarily by oxidative and defluorination pathways. Pharmacokinetic studies in humans with HFC-134a indicate rapid elimination from the body. Under acute exposure conditions, cardiac sensitization to epinephrine challenge appears to be the most sensitive end point of biological importance. The NOAELs for cardiac sensitization are 50,000, 75,000, and 250,000 ppm for HFC-134a, HFC-125, and HFC-143a, respectively. For all three components, much higher concentrations (i.e., greater than 500,000 ppm) are required to produce acute lethal or adverse clinical effects. For subchronic exposures, developmental and reproductive toxicity effects are the most sensitive indicators of biological significance. The NOAELs for developmental and reproductive effects are 2,500, 15,000, and 40,000 ppm for HFC-134a, HFC-125, and HFC-143a, respectively. However, slight maternal toxicity in the absence of other adverse effects might not be a suitable indicator for health-hazard evaluation in Navy vessels with no female crew members. Several general toxicity studies ranging from 2 to 13 weeks duration were conducted on the three components. The NOAELs for these subchronic studies are 40,000, 50,000, and 50,000 ppm for HFC-143a, HFC-134a, and HFC-125, respectively. Numerous in vitro and in vivo genotoxicity studies were performed on the three HFCs. Results indicate that none of the components are genotoxic. These genotoxicity results are in agreement with the tumor bioassays that have been performed on two of the three components. Both HFC-143a and HFC-134a showed no significant increase in the incidence of neoplasms in all organs and tissues evaluated.

Because HFC-404a is a gaseous mixture of three halocarbons (52% HFC-143a, 44% HFC-125, and 4% HFC-143a), the possibility of toxic interaction should be considered in evaluating the toxic potential of this mixture, in addition to evaluating the potential adverse health effects of the individual components. Combined exposures to multiple chemicals could result in interactions leading to a significant increase or decrease (synergism or antagonism, respectively) in overall toxicity of the mixture compared with the summation of the toxicity of individual components (Krishnan and Brodeur 1991; Mehendale 1994). However, for a large number of chemicals, the overall toxicity of a mixture can be represented by the summation of the effects of the individual components (additive effect). There are three approaches for risk assessment of chemical mixtures (Mumtaz et al. 1994). In two of these approaches, mixtures with stable composition, semi-characterized mixtures, or specially formulated mixtures are treated as a single chemical when data are available on the mixture itself. In cases when testing has been done only on the components of the mixture and not on the mixture itself, the primary method used is the hazard index (HI) approach. The HI approach is based on the principle of dose addition and uses the toxicity data available for the various components of the mixture (EPA 1986; Mumtaz et al. 1994; Teuschler and Hertzberg 1995). It is well established that some chemical components of a mixture have the potential to influence the toxicity of other components of the mixture (Groten et al. 1996). However, the HI approach does not allow the use of available interaction data. To overcome this fundamental shortcoming of the HI approach, a weight-of-evidence (WOE) method has been proposed to integrate available interaction data (Mumtaz and Durkin 1992). Recent experimental studies have shown that the WOE method is useful in assessing the toxicity of low-concentration exposures to chemical mixtures (Mumtaz et al. 1998). Because HFC-404a is an azeotrope composed of similar compounds, it is assumed that the components of the mixture would have additive effects. It is suggested that studies of HFC-404a be conducted to determine if this is a valid assumption.

EXPOSURE GUIDANCE LEVELS FOR HFC-404a

The Navy proposes to use the same exposure guidance levels for HFC-404a that were set for CFC-12 and CFC-114 (1-hr EEGL of 2,000 ppm, 24-hr EEGL of 1,000 ppm, and 90-day CEGL of 100 ppm), but did not provide an adequate rationale for doing this. To evaluate the validity of the proposed guidance levels, the subcommittee reviewed the available toxicity data on HFC-404a to determine what levels would be adequately protective of submariner health. A comparison of those results is presented below.

Submarine Exposure Guidance Levels for HFC-404a

Exposure Level	NRC's Calculated Levels	Navy's Proposed Levels
1-hr EEGL	12,900 ppm	2,000 ppm
24-hr EEGL	4,300 ppm	1,000 ppm
90-day CEGL	800 ppm	100 ppm

The subcommittee believes that the most appropriate way to calculate exposure guidance levels for HFC-404a is the method used by the American Conference of Governmental Industrial Hygienists (ACGIH 1999) to calculate Threshold Limit Values (TLVs) for special cases when the exposure of concern is a liquid mixture and the atmospheric composition is assumed to be similar to that of the original material (i.e., on a time-weighted-average exposure basis, all of the liquid mixture eventually evaporates). In that case, when the percent composition by weight of the liquid mixture is known, the exposure guidance levels can be determined using the following equation:

$$\text{TLV of mixture} = \frac{1}{\frac{f_a}{TLV_a} + \frac{f_b}{TLV_b} + \frac{f_c}{TLV_c} + \ldots \frac{f_n}{TLV_n}}.$$

The letter f stands for the fraction of each particular component. The component's corresponding TLV or, for the purposes of this report, exposure guidance level is expressed in units of milligrams per cubic meter (mg/m^3) (see Table 4-5). Using this equation, exposure guidance levels for HFC-404a were calculated using the EEGLs and CEGLs that were derived by the subcommittee for HFC-143a, HFC-125, and HFC-134a (see the following pages for calculations).

TABLE 4-5 Submarine Exposure Guidance Levels for HFC-404a Components

Exposure	Calculated Guidance Levels	
	ppm	mg/m³ᵃ
HFC-143a (molecular weight: 84.04)		
1-hr EEGL	25,000	85,930
24-hr EEGL	4,000	13,749
90-d CEGL	700	2,406
HFC-125 (molecular weight: 120.0)		
1-hr EEGL	7,500	36,810
24-hr EEGL	5,000	24,540
90-day CEGL	900	4,417
HFC-134a (molecular weight: 102.03)		
1-hr EEGL	8,000	33,384
24-hr EEGL	5,000	20,865
90-d CEGL	900	3,756

ᵃThese values were calculated using the following formula:

$$mg/m^3 = \frac{ppm \times molecular\ weight}{24.45},$$

where the value of 24.45 is the molar volume of air in liters at a pressure of 760 mm Hg and a temperature of 25°C.

The 1-hr EEGL, 24-hr EEGL, and 90-day CEGL for HFC-404a were calculated to be 12,900 ppm, 4,300 ppm, and 800 ppm, respectively. The Navy proposes to use lower guidance levels of 2,000 ppm for the 1-hr EEGL, 1,000 ppm for the 24-hr EEGL, and 100 ppm for the 90-day CEGL, which the subcommittee concludes are conservative values that are protective of submariner health.

CALCULATIONS

Calculations Used to Determine EEGLs and CEGL for HFC-404a

HFC-404a contains by weight: 52% HFC-143a, 44% HFC-125, and 4% HFC-134a

The equation below was used to calculate the 1-hr and 24-hr EEGLs and 90-

day CEGL for the mixture HFC-404a; f is the fraction of each component of the mixture and EGL is the component's corresponding exposure guidance level (which must be expressed in terms of milligrams per cubic meter):

$$\text{EGL for Mixture} = \frac{1}{\dfrac{f_a}{EGL_a} + \dfrac{f_b}{EGL_b} + \dfrac{f_c}{EGL_c} + \ldots \dfrac{f_n}{EGL_n}}.$$

Using the values in Table 4-5, the EGLs for HFC-404a were calculated as follows:

1-hr EEGL:

$$\frac{1}{\dfrac{0.52}{85{,}930 \text{ mg}/\text{m}^3} + \dfrac{0.44}{36{,}810 \text{ mg}/\text{m}^3} + \dfrac{0.04}{33{,}384 \text{ mg}/\text{m}^3}} = 52{,}075 \text{ mg}/\text{m}^3.$$

Of the mixture,

52% or 52,075 mg/m^3 × 0.52 = 27,079 mg/m^3 is HFC-143a
44% or 52,075 mg/m^3 × 0.44 = 22,913 mg/m^3 is HFC-125
4% or 52,075 mg/m^3 × 0.04 = 2,083 mg/m^3 is HFC-134a.

These values can be converted to parts per million as follows:

HFC-143a: 27,079 mg/m^3 × 0.29 = 7,853 ppm
HFC-125: 22,913 mg/m^3 × 0.20 = 4,583 ppm
HFC-134a: 2,083 mg/m^3 × 0.24 = 500 ppm.

The 1-hr EEGL of HFC-404a = 7,853 + 4,583 + 500 ≈ 12,900 ppm.

24-hr EEGL:

$$\frac{1}{\dfrac{0.52}{13{,}749 \text{ mg}/\text{m}^3} + \dfrac{0.44}{24{,}540 \text{ mg}/\text{m}^3} + \dfrac{0.04}{20{,}865 \text{ mg}/\text{m}^3}} = 17{,}341 \text{ mg}/\text{m}^3.$$

Of the mixture,

52% or 17,341 mg/m^3 × 0.52 = 9,017 mg/m^3 is HFC-143a
44% or 17,341 mg/m^3 × 0.44 = 7,630 mg/m^3 is HFC-125
4% or 17,341 mg/m^3 × 0.04 = 694 mg/m^3 is HFC-134a.

These values can be converted to parts per million as follows:

HFC-143a: 9,017 mg/m³ × 0.29 = 2,615 ppm
HFC-125: 7,630 mg/m³ × 0.20 = 1,526 ppm
HFC-134a: 694 mg/m³ × 0.24 = 167 ppm.

24-hr EEGL of HFC-404a = 2,615 + 1,526 + 167 ≈ 4,300 ppm.

90-day CEGL:

$$\frac{1}{\frac{0.52}{2,406 \text{ mg/m}^3} + \frac{0.44}{4,417 \text{ mg/m}^3} + \frac{0.04}{3,756 \text{ mg/m}^3}} = 3,064 \text{ mg/m}^3.$$

Of the mixture,

52% or 3,064 mg/m³ × 0.52 = 1,593 mg/m³ is HFC-143a
44% or 3,064 mg/m³ × 0.44 = 1,348 mg/m³ is HFC-125
 4% or 3,064 mg/m³ × 0.04 = 123 mg/m³ is HFC-134a.

These values can be converted to parts per million as follows:

HFC-143a: 1,593 mg/m³ × 0.29 = 462 ppm
HFC-125: 1,348 mg/m³ × 0.20 = 270 ppm
HFC-134a: 123 mg/m³ × 0.24 = 30 ppm.

90-day CEGL of HFC-404a = 462 + 270 + 30 = 800 ppm.

REFERENCES

ACGIH (American Conference of Governmental Industrial Hygienists). 1999. TLVs and BEIs. Threshold Limit Values for Chemical Substances and Physical Agents. Biological Exposure Indices. Cincinnati, OH.: American Conference of Governmental Industrial Hygienists.

AIHA (American Industrial Hygiene Association). 1991. Workplace Environmental Exposure Level Guide: 1,1,1,2-Tetrafluoroethane. Akron, OH.: American Industrial Hygiene Association.

AIHA (American Industrial Hygiene Association). 1996. Workplace Environmental Exposure Level: 1,1,1-Trifluoroethane. Fairfax, Va.: American Industrial Hygiene Association.

Alexander, D.J., and S.E. Libretto. 1995. An overview of the toxicology of HFA-134a (1,1,1,2-tetrafuloeoethane). Human Exper. Toxicol. 14:715-720.

Alexander, D.J., E. Mortimer, G.D. Dines, S.E. Libretto, and D.N. Mallett. 1995. One-year study in dogs of the toxicity of HFA-134a by inhalation. Inhalation Toxicol. 7:1153-1162.

Alexander, D.J., S.E. Libretto, M.J. Adams, E.W. Hughes, and M. Bannerman. 1996. HFA-134a (1,1,1,2-tetrafluoroethane): effects of inhalation exposure upon reproductive performance, development and maturation of rats. Human Exp. Toxicol. 15:508-517.

Anderson, D., and C.R. Richardson. 1979. Arcton 134a: A Cytogenetic Study in the Rat. Study Number SR0002, Report Number CTL/P/444, Central Toxicology Laboratoy, ICI, England.

Araki, A. 1991. Report on Reverse Mutation Assay in Bacteria on Tetrafluoroethane. Japan Bioassay Laboratory, Study Nos. 5292 & 5312. Japan Industrial Safety and Health Association.

Asakura, M. 1991. Report on a Chromosomal Aberration Test of 1,1,1,2-Tetrafluoroethane in Cultured Mammalian Cells. Japan Bioassay Laboratory, Study Number 5879. Japan Industrial Safety and Health Association.

Azar, A., H.J. Trochimowicz, J.B. Terrill, and L.S. Mullin. 1973. Blood levels of fluorocarbon related to cardiac sensitization. Am. Ind. Hyg. Assoc. J. 34:102-109.

Barton, S.J., P. McDonald, and J. Sandow. 1994. HFA-134a Study of the Effects on Testicular Endocrine Function After Inhalation Exposure (6 h per day). IRI Project No. 490704. Study prepared by Inveresk Research International, Tranent, Scotland, for International Pharmaceutical Aerosol Consortium for Toxicology Testing, Washington, D.C.

Brock, W.J., H.J. Trochimowicz, C.H. Farr, R.J. Millischer, and G.M. Rusch. 1996. Acute, subchronic, and developmental toxicity and genotoxicity of 1,1,1-trifluoroethane (HFC-143a). Fundam. Appl. Toxicol. 31:200-209.

Brooker, A.J., P.J. Brown, D.M. John, and D.W. Coombs. 1992. The Effect of HFC 125 on Pregnancy of the Rabbit. Report No. ALS 10/920856. Huntingdon Research Centre, Cambridgeshire, England.

Brusick, D.J. 1976. Mutagenicity data of Genetron 134a. Final Report, LBI Report No. 2683, (unpublished data), Litton Bionetics.

Callander, R.D., and K.P. Priestley. 1990. HFC 134a. An Evaluation Using the Salmonella Mutagenicity Assay. Report No. CTL/P/2422, Central Toxicology Laboratory,(unpublished data), ICI, England.

Collins, M.A. 1984. HFC 134a: Acute Toxicity in Rats to Tetrafluoroethane. Unpublished data from Central Toxicology Laboratory, ICI, England.

Collins, M.A., G.M. Rusch, F. Sato, P.M. Hext, and R.J. Millischer. 1995. 1,1,1,2-Tetrafluoroethane: repeat exposure inhalation toxicity in the rat, developmental toxicity in the rabbit, and genotoxicity in vitro and in vivo. Fundam Appl. Toxicol. 25:271-280.

Dance, C.A., and G. Hodson-Walker. 1992. In Vitro Assessment of the Clastogenic

Activity of HFC 125 in Cultured Chinese Hamster Ovary (CHO) Cells. LSR Report No. 91/PAR006/1015a. Life Science Research Ltd., Eye, Suffolk, England.

Dance, C.A., K.E. Beach, and G. Hodson-Walker. 1992. In Vitro Assessment of the Clastogenic Activity of HFC 125 in Cultured Human Lymphocytes. LSR Report No. 91/PAR005/1014a. Life Science Research Ltd., Eye, Suffolk, England.

DuPont Company. 1994. Metabolism of HFC-143a in the Rat. Haskell Laboratory Report No. 3-94. (Unpublished data.) Haskell Laboratory, E.I. du Pont de Nemours & Co., Newark, DE.

ECETOC (European Centre for Ecotoxicology and Toxicology of Chemicals). 1994. Joint Assessment of Commodity Chemicals No. 24, Pentafluoroethane (HFC-125), May, 1994. [ISSN-0773-6339-24].

ECETOC (European Centre for Ecotoxicology and Toxicology of Chemicals). 1995. Joint Assessment of Commodity Chemicals No. 31, 1,1,1,2-Tetrafluoroethane (HFC-134a), February, 1995. [ISSN-0773-6339-31].

Edwards, C.N., G. Hodson-Walker, and S. Cracknell. 1992. HFC 125: Assessment of Clastogenic Action on Bone Marrow Erythrocytes in the Micronucleus Test. LSR Report No. 92/PAR004/0148. Life Science Research Ltd., Eye, Suffolk, England.

Ellis, M.K., L.A. Gowans, T. Green, and R.J.N. Tanner. 1993. Metabolic fate and disposition of 1,1,1,2-tetrafluoroethane (HFC134a) in rat following a single exposure by inhalation. Xenobiotica 23:719-729.

Emmen, H.H., and E.M.G. Hoogendijk. 1999. Report on an Ascending Dose Safety Study Comparing HFA-134a with CFC-12 and Air, Administered by Whole-Body Exposure to Healthy Volunteers. TNO Report V98.754 – Vol. 1 and 2, Zeist, The Netherlands: TNO Nutrition and Food Research Institute.

EPA (U.S. Environmental Protection Agency). 1986. Guidelines for Health Risk Assessment of Chemical Mixtures. 51 FR 34014. Sept, 24.

Finch, J.R., E.J. Dadey, S.L. Smith, L.I. Harrison, and G.A. Digenis. 1995a. Dynamic monitoring of total-body absorption by ^{19}F NMR spectroscopy: one hour ventilation of HFA-134a in male and female rats. Magn. Reson. Med. 33:409-413.

Finch, J.R., W.R. Banks, D.R. Hwang, M.R. Satter, B. Ezzidene, J.C. Mantil, and G.A. Digenis. 1995b. Synthesis and in vivo disposition studies of ^{18}F-labeled HFA-134a. Appl. Radiat. Isot. 46:241-248.

Groten J.P., E.D. Schoen, P.J. van Bladeren, C.F. Kuper, J.A. van Zorge, and V.J. Feron. 1996. Subacute toxicity of a mixture of nine chemicals in rats: detecting interactive effects with a fractionated two-level factorial design. Fundam. Appl. Toxicol. 36:15-29.

Hardy, C.J. 1992. Assessment of Cardiac Sensitization Potential in Dogs: Comparison of HFC-125 and Halon 13B1 Report No. ALS11/920116. Huntingdon Research Centre Ltd., Cambridgeshire, England.

Hardy, C.J., I.J. Sharman, and G.C. Clark. 1991. Assessment of Cardiac Sensitization Potential in Dogs. Rep. No. CTL/C/2521. Huntingdon Research Centre, Cambridgeshire, U.K.

Harris, J.W., J.P. Jones, J.L. Martin, A.C. La Rosa, M.J. Olson, L.R. Pohl, and M.W. Anders. 1992. Pentahaloethane-based chlorofluorocarbon substitutes and halothane: correlation of in vivo hepatic protein-trifluoroacetylation and urinary trifluoroacetic acid excretion with calculated enthalpies of activation. Chem. Res. Toxicol. 5:720-725.

Harrison, L.I., D. Donnell, J.L. Simmons, B.P. Ekholm, K.M. Cooper, and P.J. Wyld. 1996. Twenty-eight-day double-blind safety study of an HFA-134a inhalation aerosol system in healthy subjects. J. Pharm. Pharmacol. 48:596-600.

Hext, P.M. 1989. HFC-134a: 90-Day Inhalation Toxicity Study in the Rat. ICI Rep. No. CTL/P/2466. Central Toxicology Laboratory, Imperial Chemical Industries, Alderley Park, Macclesfield, Cheshire, U.K.

Hext, P.M., and R.J. Parr-Dobrzanski. 1993. HFC-134a: A 2-Year Inhalation Toxicity Study in the Rat. ICI Rep. No. CTL/P/3841. Central Toxicology Laboratory, Imperial Chemical Industries, Alderley Park, Macclesfield, Cheshire, U.K.

Hodge, M.C.E., M. Kilmartin, R.A. Riley, T.M. Weight, and J. Wilson. 1979a. Arcton 134a: Teratogenicity Study in the Rat. ICI Rep. No. CTL/P/417. Central Toxicology Laboratory, Imperial Chemical Industries, Alderley Park, Macclesfield, Cheshire, U.K.

Hodge, M.C.E., D. Anderson, I.P. Bennett, and T.M. Weight. 1979b. Arcton 134a: Dominant Lethal Study in the Mouse. ICI Rep. No. CTL/P/437, Central Toxicology Laboratory, Imperial Chemical Industries, Alderley Park, Macclesfield, Cheshire, U.K.

ICI Chemicals & Polymers. 1991. Cardiac Sensitization Study in Dogs. [Personal communication]. I.C.I. Chemicals & Polymers, Attn: Dr. David Farrar. P.O. Box 14, The Heath , Runcorn, Cheshire WA7 4QG, England.

Kawano, T., H.J. Trochimowicz, G. Malinverno, and G.M. Rusch. 1995. Toxicological evaluation of 1,1,1,2,2-pentafluororethane (HFC-125). Fundam. Appl. Toxicol. 28:223-231.

Krishnan, K., and J. Brodeur. 1991. Toxicological consequences of combined exposures to environmental pollutants. Arch. Complex Environ. Studies 3:1-106.

Longstaff, E., M. Robinson, C. Bradbrook, J.A. Styles, and I.F.H. Purchase. 1984. Genotoxicity and carcinogenicity of fluorocarbons: assessment by short-term in vitro tests and chronic exposure in rats. Toxicol. Appl. Pharmacol. 72:15-31.

Lu, M., and R. Staples. 1981. 1,1,1,2-Tetrafluoroethane (FC-134a): Embryo-Fetal Toxicity and Teratogenicity Study by Inhalation in the Rat. Report No. 317-81. Haskell Laboratory, Wilmington, DE.

Mackay, J.M. 1990. HFC 134a. An Evaluation in the in Vitro Cytogenetic Assay in Human Lymphocytes. Report Number CTL/P/2977, Central Toxicology Laboratory, ICI, England.

Master, R.E., R.J. Brown, D.M. John, and D.W. Coombs. 1992. A Study of the Effect of HFC 125 on Pregnancy of the Rat (Inhalation Exposure). Report No. ALS 9/920434. Huntingdon Research Centre Ltd., Cambridgeshire, England.

May, K., D. Watson, and G. Hodson-Walker. 1992. HFC 125 in Gaseous Phase: Assessment of Mutagenic Potential in Amino Acid Auxotrophs of Salmonella

Typhimurium and Escherichia Coli (the Ames Test). LSR Report No. 91/PAR003/1152a. Life Science Research Ltd., Eye, Suffolk, England.
Mehendale, H.M. 1994. Amplified interactive toxicity of chemicals at nontoxic levels: mechanistic considerations and implications to public health. Environ. Health Perspect. 102 (Suppl. 9):139-149.
Mercier, O. 1989. HFA-134a: Test to Determine the Index of Primary Cutaneous Irritation in the Rabbit. Report No. 911422. Hazelton, France.
Mercier, O. 1990a. HFA-134a: Test to Evaluate the Ocular Irritation in the Rabbit. Report No. 912349. Hazelton, France.
Mercier, O. 1990b. HFA-134a: Test to Evaluate the Sensitising Potential by Topical Applications in the Guinea Pig. The Epicutaneous Maximisation Test. Report No. 001380. Hazelton, France.
Monte, S.Y., I. Ismail, D.N. Mallett, C. Matthews, and R.J. Tanner. 1994. The minimal metabolism of inhaled 1,1,1,2-tetrafluoroethane to trifluoroacetic acid in man as determined by high sensitivity ^{19}F nuclear magnetic resonance spectroscopy of urine samples. J. Pharm. Biomed. Anal. 12:1489-1493.
Müller, W., and T. Hofmann. 1989. CFC 134a: Micronucleus Test in Male and Female NMRI Mice After Inhalation. Study Number 88.1244. Pharma Research Toxicology and Pathology, Hoechst AG, Federal Republic of Germany.
Mullin, L.S., and R.W. Hartgrove. 1979. Cardiac Sensitization. Rep. No. 42-79. Haskell Laboratory, Wilimington, DE.
Mullin, L.S., C.F. Reinhardt, and R.E. Hemingway. 1979. Cardiac arrhythmias and blood levels associated with inhalation of Halon 1301. Am. Ind. Hyg. Assoc. J. 40:653-658.
Mumtaz, M.M., and P.R. Durkin. 1992. A weight-of-evidence approach for assessing interactions in chemical mixtures. Toxicol. Ind. Health 8:377-406.
Mumtaz, M.M., C.T. DeRosa, and P.R. Durkin. 1994. Approaches and challenges in risk assessments of chemical mixtures. Pp. 565-597. In: Toxicology of Chemical Mixtures: Case Studies, Mechanisms, and Novel Approaches, R.S.H. Yang, ed., San Diego, CA.: Academic Press.
Mumtaz, M.M., C.T. De Rosa, J. Groten, V.J. Feron, H. Hansen, and P.R. Durkin. 1998. Estimation of toxicity of chemical mixtures through modeling of chemical interactions. Environ Health Perspect. 106(suppl 6.): 1353-1360.
NRC (National Research Council). 1996. Toxicity of Alternatives to Chlorofluorocarbons: HFC-134a and HCFC-123. Washington D.C.: National Academy Press.
Nakayama, E., K. Nagano, M. Ohnishi, S. Katagiri, and O. Montegi. 1992a. Acute Inhalation Toxicity Study of 1,1,1,2,2-Pentafluoroethane in Rats. Study No. 0184. Japan Bioassay Laboratory, Hirasawa, Japan.
Nakayama, E., K. Nagano, M. Ohnishi, and O. Montegi. 1992b. Four-week Inhalation Study of 1,1,1,2,2-Pentafluoroethane (HFC 125) in Rats. Study No. 0182. Japan Bioassay Laboratory, Hirasawa, Japan.
Nakayama, E., K. Nagano, M. Ohnishi, and O. Montegi. 1993. Thirteen-week Inhalation Toxicity Study of 1,1,1,2,2-Pentafluoroethane (HFC 125) in Rats. Study No. 0197, Japan Bioassay Laboratory, Hirasawa, Japan.

Nikijenko, T.K., and M.S. Tolgskaya. 1965. On the toxico-pathomorphological changes in animals under the effect of Freons-141, 142, and 143, and the intermediate products of their production. [Russian]. Gig. Tr. Prof. Zabol. 9:37-44.

Olson, M.J., and S.E. Surbrook, Jr. 1991. Defluorination of the CFC-substitute 1,1,1,2-tetrafluoroethane: comparison in human, rat and rabbit hepatic microsomes. Toxicol. Lett. 59:89-99.

Olson, M.J., C.A. Reidy, and J.T. Johnson. 1990a. Defluorination of 1,1,1,2-tetrafluoroethane (R-134a) by rat hepatocytes. Biochem. Biophys. Res. Commun. 166:1390-1397.

Olson, M.J., C.A. Reidy, J.T. Johnson, and T.C. Pederson. 1990b. Oxidative defluorination of 1,1,1,2-tetrafluoroethane (R-134a) by rat liver microsomes. Drug Metab. Dispos. 18:992-998.

Olson, M.J., S.G. Kim, C.A. Reidy, J.T. Johnson, and R.F. Novak. 1991. Oxidation of 1,1,1,2-tetrafluoroethane in rat liver microsomes is catalyzed primarily by cytochrome P450IIE1. Drug Metab. Dispos. 19:298-303.

PAFT (Programme for Alternative Toxicology Testing). 1989. Toxicology Forum, European Symposium, Toulouse, France.

Panepinto, A.S. 1990. Four Hours Inhalation Approximate Lethal Concentrations (ALC) of HFC 125, Haskell Laboratory Report 582-90, Haskell Laboratory, DuPont.

Pike, V.W., F.I. Aigbirhio, C.A.J. Freemantle, B.C. Page, C.G. Rhodes, S.L. Waters, T. Jones, P. Olsson, G.P. Ventresca, R.J.N. Tanner, M. Hayes, and J.M.B. Hughes. 1995. Disposition of inhaled 1,1,1,2-tetrafluoroethane (HFA134A) in healthy subjects and in patients with chronic airflow limitation. Measurement by ^{18}F-labeling and whole-body γ-counting. Drug Metab. Dispos. 23:832-839.

Riley, R.A., I.P. Bennet, I.S. Chart, C.W. Gore, M. Robinson, and T.M. Weight. 1979. Arcton 134a: Subacute Toxicity to the Rat by Inhalation. ICI Rep. No. CTL/P/463, Central Toxicology Laboratory, Imperial Chemical Industries, Alderley Park, Macclesfield, Cheshire, U.K.

Rissolo, S.B., and J.A. Zapp. 1967. Acute Inhalation Toxicity. Rep. No. 190-67, Haskell Laboratory, Wilmington, DE.

Silber, L.S., and G.L. Kennedy. 1979a. Acute Inhalation Toxicity of Tetrafluoroethane. Rep. No. 422-79. Haskell Laboratory, Wilmington, DE.

Silber, L.S., and G.L. Kennedy. 1979b. Subacute Inhalation Toxicity of Tetrafluoroethane (FC-134a). Rep. No. 228-79. Haskell Laboratory, Wilmington, DE.

Surbrook, S.E., Jr., and M.J. Olson. 1992. Dominant role of cytochrome P-450 2E1 in human hepatic microsomal oxidation of the CFC-substitute 1,1,1,2-tetrafluoroethane. Drug Metab. Dispos. 20:518-524.

Teuschler, L.K., and R.C. Hertzberg. 1995. Current and future risk assessment guidelines, policy, and methods development for chemical mixtures. Toxicology 105:137-144.

Trochimowicz, H.J., A. Azar, J.B. Terrill, and L.S. Mullin. 1974. Blood levels of fluorocarbon related to cardiac sensitization: Part II. Am. Ind. Hyg. Assoc. J. 35:632-639.

Trueman, R.W. 1990. HFC 134a: Assessment for the Induction of Unscheduled DNA

Synthesis in Rat Hepatocytes in Vivo. Study Number SR0337, Report No. CTL/P/2550, Central Toxicology Laboratory, ICI, England.

Van Demark, N.L., and M.J. Fre. 1970. Temperature ffects. Pp. 235-245. In: Testis, A.D. Johnson, W.R. Gores, and N.L. Van Demark, Eds. New York, NY: Academic Press.

Vinegar, A., and G.W. Jepson. 1995. Relating Blood Concentration Time Courses to Cardiac Sensitization Thresholds During Inhalation of Halon Replacement Chemicals. Report No. AL/OE-TR-1995-0132. U.S. Air Force, Armstrong Laboratory, Wright-Patterson Air Force Base, OH.

Vinegar, A., R.S. Cook, J.D. McCafferty, III, M.C. Caracci, and G.W. Jepson. 1997. Human Inhalation of Halon 1301, HFC-134a and HFC-227ea for Collection of Pharmacokinetic Data. Interim Report No. AL/OE-TR-1997-0116, U.S. Air Force, Armstrong Laboratory, Wright-Patterson Air Force Base, OH.

Wang, Y., M.J. Olson, and M.T. Baker. 1993. Interaction of fluoroethane chlorofluorocarbon (CFC) substitutes with microsomal cytochrome P450. Stimulation of P450 activity and chlorodifluoroethene metabolism. Biochem. Pharmacol. 46:87-94.

Wickramaratne, G.A. 1989a. HFC-134a: Teratogenicity Inhalation Study in the Rabbit. ICI Rep. No. CTL/P/2504. Central Toxicology Laboratory, Imperial Chemical Industries, Alderley Park, Macclesfield, Cheshire, U.K.

Wickramaratne, G.A. 1989b. HFC-134a: Embryotoxicity Inhalation Study in the Rabbit. ICI Rep. No. CTL/P/2380. Central Toxicology Laboratory, Imperial Chemical Industries, Alderley Park, Macclesfiled, Cheshire, U.K.